Circuit Simulation
and Analysis
by Dr. Saeid Moslehpour

An introduction to computer-aided circuit design using *PSpice* software

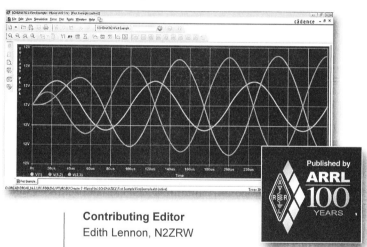

Published by
ARRL
100 YEARS

Contributing Editor
Edith Lennon, N2ZRW

Production
Shelly Bloom, WB1ENT
Jodi Morin, KA1JPA
Maty Weinberg, KB1EIB
David Pingree, N1NAS

Cover Design
Sue Fagan, KB1OKW

Contents

Foreword

The circuit design process is usually interactive, with testing along the way to ensure that a circuit performs as expected, followed by design changes and more testing. The time-honored way to do this was to build a physical prototype, test it over a range of range of operating voltages, signal levels, temperatures and other factors, tweak it and test some more.

In the 1970s, scientists at the University of California, Berkeley, developed *Simulated Program with Integrated Circuit Emphasis (SPICE)*, a program for simulating circuits on a mainframe computer. Now engineers could create, analyze and modify circuits using software, coming much closer to the final design before soldering a single component.

Fast-forward to today, and *SPICE,* along with its derivatives, is hugely popular for predicting the behavior of electronic circuits. *PSpice* from Cadence Design Systems is a version of *SPICE* that runs on personal computers and offers many features, component libraries and tools for circuit designers. Anyone performing circuit analysis or design should have a working knowledge of *PSpice* in order to save time and gain insight into circuit behavior by answering "what if" questions with a computer simulation. Students, professionals and hobbyists already familiar with traditional methods of circuit analysis will find *PSpice* to be an important tool for learning circuit analysis and design and for testing electronic circuits in ways they could not easily do in many laboratories.

In the book, Dr Saeid Moslehpour introduces readers to the basic functions and tools needed to create simple circuits and analyze their behavior using *PSpice*. To get the most from this book, readers are expected to obtain a version of this software (see Chapter 1), and follow along with the examples.

David Sumner, K1ZZ
Chief Executive Officer
Newington, Connecticut
October 2013

Preface

The first two editions of this book were published in 1988 and 1990 by my colleague, mentor and friend, Professor Walter Banzhaf, WB1ANE, and it was admired by many in the electrical engineering field. I had this idea of upgrading the book back in 2008, and after five years all those efforts paid off. What you have here is almost a new book.

PSpice was initially developed by MicroSim and was used in electronic design automation. The company was bought by OrCAD, which was subsequently purchased by Cadence Design Systems. *PSpice* was the first version of UC Berkeley *SPICE* available on a PC, having been released in January 1984 to run on the original IBM PC. It remains one of the best known and most widely used packages on the market today.

I am using Cadence Design Systems version 16.5 for all simulations in this book. Although there are many modules within this software package, I am emphasizing use of the tools available in *OrCAD Capture* and *PSpice* (or in the full package from Cadence Design systems, *Allegro Design Entry Capture* and *Allegro AMS Simulator*). More information on this software is found in Chapter 1.

I express deep appreciation to Walter Banzhaf who created 1st and 2nd editions, and my handling editor Edith Lennon, N2ZRW. I also thank the talented audio engineer Brandon LaChance, my former student and great engineer, who helped with many of the simulations and without whom this book couldn't have been published. A special word of thanks to my wife, Parisa, and my mother Giti, and my father, Amir.

Finally I would like to thank you for choosing this book. I hope this book will enhance your knowledge of electrical/electronic design and simulation. Please share your suggestions and comments. There is a feedback form at the back of the book for this purpose, or send an e-mail to pubsfdbk@arrl.org.

Saeid Moslehpour, PhD
Professor of Electrical Engineering,
  University of Hartford
West Hartford, Connecticut
October 2013

About the Author

Saeid Moslehpour, PhD, is Associate Professor of Electrical and Computer Engineering and Department Chair at the University of Hartford in Connecticut. Dr Moslehpour holds BS and MS degrees in Electronics from the University of Central Missouri and a PhD in Industrial Technology and Computer Engineering from Iowa State University. He is a Senior Master Engineer in the National Association of Radio and Telecommunications Engineers, a Senior Member, IEEE and Member, ASEE.

About the ARRL

The seed for Amateur Radio was planted in the 1890s, when Guglielmo Marconi began his experiments in wireless telegraphy. Soon he was joined by dozens, then hundreds, of others who were enthusiastic about sending and receiving messages through the air—some with a commercial interest, but others solely out of a love for this new communications medium. The United States government began licensing Amateur Radio operators in 1912.

By 1914, there were thousands of Amateur Radio operators—hams—in the United States. Hiram Percy Maxim, a leading Hartford, Connecticut inventor and industrialist, saw the need for an organization to band together this fledgling group of radio experimenters. In May 1914 he founded the American Radio Relay League (ARRL) to meet that need.

Today ARRL, with approximately 155,000 members, is the largest organization of radio amateurs in the United States. The ARRL is a not-for-profit organization that:

- promotes interest in Amateur Radio communications and experimentation
- represents US radio amateurs in legislative matters, and
- maintains fraternalism and a high standard of conduct among Amateur Radio operators.

At ARRL headquarters in the Hartford suburb of Newington, the staff helps serve the needs of members. ARRL is also International Secretariat for the International Amateur Radio Union, which is made up of similar societies in 150 countries around the world.

ARRL publishes the monthly journal *QST* and an interactive digital version of *QST*, as well as newsletters and many publications covering all aspects of Amateur Radio. Its headquarters station, W1AW, transmits bulletins of interest to radio amateurs and Morse code practice sessions. The ARRL also coordinates an extensive field organization, which includes volunteers who provide technical information and other support services for radio amateurs as well as communications for public-service activities. In addition, ARRL represents US amateurs with the Federal Communications Commission and other government agencies in the US and abroad.

Membership in ARRL means much more than receiving *QST* each month. In addition to the services already described, ARRL offers membership services on a personal level, such as the Technical Information Service—where members can get answers by phone, email or the ARRL website, to all their technical and operating questions.

Full ARRL membership (available only to licensed radio amateurs) gives you a voice in how the affairs of the organization are governed. ARRL policy is set by a Board of Directors (one from each of 15 Divisions). Each year, one-third of the ARRL Board of Directors stands for election by the full members they represent. The day-to-day operation of ARRL HQ is managed by an Executive Vice President and his staff.

No matter what aspect of Amateur Radio attracts you, ARRL membership is relevant and important. There would be no Amateur Radio as we know it today were it not for the ARRL. We would be happy to welcome you as a member! (An Amateur Radio license is not required for Associate Membership.) For more information about ARRL and answers to any questions you may have about Amateur Radio, write or call:

ARRL—the national association for Amateur Radio®
225 Main Street
Newington CT 06111-1494
Voice: 860-594-0200
Fax: 860-594-0259
E-mail: **hq@arrl.org**
Internet: **www.arrl.org**

Prospective new amateurs call (toll-free):
800-32-NEW HAM (800-326-3942)
You can also contact us via e-mail at **newham@arrl.org**
or check out the ARRL website at **www.arrl.org**

Chapter 1

Introduction

Circuit analysis is a necessary part of circuit design. Once a design for a circuit has been determined, the soundness of that design must be tested to ensure that the circuit does indeed perform as required and intended. Often this involves testing for dc operating point and performance, with signal applied, over a range of dc supply voltages, input signal levels and temperatures. The time-honored way to do this was to build a prototype of the circuit, send it off to the laboratory, and invest large amounts of time and money in putting the circuit through its paces in the hope that it would perform as desired.

The Need for Circuit Analysis

A much better method was introduced with *Simulated Program with Integrated Circuit Emphasis* (*SPICE*), a software program developed in the 1970s to allow circuits to be simulated on a mainframe computer, thus saving huge amounts of time and money in the early development stages of circuit design. Using this software would allow electrical engineers to reduce lengthy calculations needed for electrical/electronic circuit design. Engineers would design the circuits and view the results in texts or graphical format. Cadence Design Systems, Inc (**www.cadence.com**) promotes the same software under two names: Cadence SPB (**www.cadence.com/products/orcad/Pages/default.aspx**) and OrCAD (**www.ema-eda.com**); they are essentially same products with two different names.

SPICE is an early predecessor of the Cadence *PSpice* (*Personal Simulation Program with Integrated Circuit Emphasis*) software explored in this book.

To appreciate the capabilities and continual development of *PSpice* it is helpful to have an understanding of its origins, but one does not need to understand the algorithms and mathematical techniques that *PSpice* uses to analyze circuits in order to successfully solve circuit problems with *PSpice*, any more than one needs a working knowledge of internal combustion engines to drive a car. However, it is vitally important for anyone using *SPICE* (or any other computer analysis/design program) to be a competent practitioner in the field of electrical and computer engineering. Only in that way can one check the correctness and reasonableness of the computer's answers by estimation using sound engineering judgment and/or by working a sample problem through by conventional means.

PSpice Overview

PSpice is a PC version of *SPICE* that runs on personal computers and offers many improvements over other releases. *SPICE,* along with its derivatives, is the most popular computer program in the world today for predicting the behavior of electronic circuits. It was developed by the Integrated Circuits Group of the Electronics Research Laboratory and the Department of Electrical Engineering and Computer Sciences at the University of California, Berkeley, California. The person credited with originally developing *SPICE* is Dr. Lawrence Nagel, whose PhD thesis describes the algorithms and numerical methods used in *SPICE*. The software has undergone many changes since it was first developed, and it continues to evolve.

SPICE is a large (over 17,000 lines of Fortran source code), powerful, and extremely versatile industry-standard program for circuit analysis and IC design. A significant number of companies have customized Berkeley's *SPICE* for in-house circuit development work. Many software packages based on *SPICE* have been developed which use the *SPICE2* program, also from Berkeley, as the core for performing circuit analysis. Most of these have added useful programs to make the complete package easier to use. For example, *SPICE2* is not interactive, it does not have the capability to reference a library of semiconductor components, and its graphs are made on a line printer using American Standard Code for Information Interchange (ASCII) symbols. Although it did not feature interactive libraries, it enabled the user to create models for metal-oxide-semiconductor field-effect transistors (MOSFETs), bipolar junction transistors (BJTs), field-effect transistors (FETs), and diodes. Some of the commercially available packages are interactive, include extensive libraries of parts, and have graphics post-processors that make professional-looking graphs.

SPICE3 was developed in the late 1980s, modified, and rewritten using the general purpose programming language C. The first stable release of *SPICE3*, in 1993, featured advanced component models for MOSFETs. Initially, *SPICE* relied on a mainframe-based operation and has since been modified to function on some of the more common operating systems and personal computers. With this innovation, the acronym has been amended from *SPICE* to *PSpice*, the P, of course, standing for personal.

At the time, *OrCAD*, short for *Oregon Computer Aided Design*, was becoming increasingly popular in the design industry for its ability to create schematic representations of designs and printed circuit board layouts. Founded in Hillsboro, Oregon, in 1985 by John Durbetaki and Ken and Keith Seymour as OrCAD Systems Corporation, the company was acquired in the late 1990s by Cadence Design Systems to be used in conjunction with the advanced *PSpice* simulators to function as a schematic capture software program to interpret a schematic design input by a user for simulation purposes. This is vastly different from some of the primitive versions of *SPICE* that required the user to enter a schematic design into the software by means of a *netlist*, a text representation of a schematic derived by the user.

As of this writing, the most recent release of the *OrCAD/PSpice* software is version 16.6.[1] *PSpice*, and other packages based on it, will be around and widely used for many years to come. As you read this book, you will see that *PSpice*, though developed for the design of integrated circuits, can be used to solve a great variety of non-IC circuit problems involving power supplies, three-phase power systems, transmission lines, and nonlinear components, to name a few applications. Anyone performing circuit analysis or design should have a working knowledge of *PSpice* in order to save time and money and to gain insight into circuit behavior by answering "what if" questions with a computer simulation. "What if" questions are frequently not answered if the tedium involved in doing so outweighs the curiosity of the person asking the question.

Students, professionals and hobbyists will find *PSpice* to be an important tool for learning circuit analysis and design and for testing electronic circuits in ways they could not easily do in many laboratories. By learning a version of *SPICE*, users will be preparing for the kind of circuit simulation they will encounter in industry. They should, however, be competent in traditional methods of circuit analysis before embarking on computer methods.

A Word on Nomenclature

Several different packages of the Cadence design and analysis software are available and were used in preparation of this book. At the time the book was written, Version 16.5 was most current, but version 16.6 is now available. Regardless of where it was obtained, the software packages are identical and features differ only in name. In later chapters of this book you will see examples created with the various packages.

EMA Design Automation

If the software is obtained from EMA Design Automation (**www.ema-eda.com**), the software will be called:

Schematic capture — *OrCAD Capture* or *OrCAD Capture CIS*
Circuit simulation — *PSpice A/D* or *PSpice Advanced Analysis*

Cadence Design Systems

If the software is obtained from Cadence Design Systems (**www.cadence.com**), the software will be called:

Schematic capture — Cadence *Allegro Design Entry Capture* or *Allegro Design Entry Capture CIS*
Circuit simulation — Cadence *Allegro AMS Simulator*

Demo Version

Demo versions are available for download from the Cadence website at **www.cadence.com/products/orcad/pages/downloads.aspx**. The demo package includes *OrCAD Capture CIS Lite* and *PSpice A/D Lite*, which are fully functional versions that can be used to explore the examples in this book. There are some limitations on circuit and analysis complexity, but these are very useful and powerful tools.

Support Files

Modeling files for some of the examples described in this book are available from the ARRL website at **www.arrl.org/circuit-simulation**.

Notes
[1]Source for company history and development, **www.fundinguniverse.com/company-histories/cadence-design-systems-inc-history/**.

Chapter 2

The *OrCAD* User Interface

In the last chapter, the main topics of discussion were an overview of the Cadence software. You were also provided with a brief history, an outline of the different incarnations of the Cadence/*PSpice* software, and a few general statements giving an overview of its capabilities.

Familiarization and Preparation of the *Design Entry* User Interface

In this chapter, you will become familiar with the graphical user interfaces (GUIs) utilized in the Cadence software, one of which is the schematic capture tool, *Cadence Allegro Design Entry*, and the other is the *Cadence Allegro AMS Simulator*. (In Chapter 1 we learned that these products are called *OrCAD Capture* and *PSpice A/D* in the EMA Design version and *OrCAD Capture CIS Lite* and *PSpice A/D Lite* in the demo version. We'll refer to them as *Design Entry* and *AMS Simulator*, but the screens look and work the same for either version.) By looking at three things — the circuit schematic diagram, the simulation settings, and the output file — you will gain an appreciation of what the Cadence software can do and start to use it yourself. A good way to begin learning *PSpice* is to create input files via *Design Entry* schematics similar to the ones in this chapter and then run analyses of them.

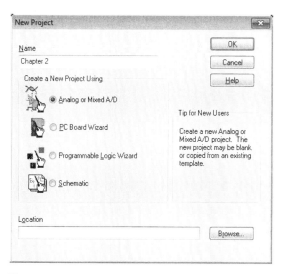

Figure 2.1 — NEW PROJECT window.

Before beginning any type of simulation using the *Design Entry* and *AMS Simulator*, you must create a blank project. Upon selecting the *Design Entry* software icon, you will be presented with the window shown in **Figure 2.1**.

This window allows the user to designate a name for the project and select a directory in which the project can be stored. In order to carry out the basic simulations covered in this text, it is important to select the button labeled ANALOG OR MIXED A/D. A common mistake made by new users of the Cadence software is the selection of the SCHEMATIC file type. This schematic file type will only allow the user to create a schematic design of a circuit, but will not allow a netlist or simulations to be generated.

The project used to generate the schematics and simulations that illustrate this chapter has been titled, "Chapter2." When creating your own projects, you may store them to a destination of your choosing by either entering a directory manually in the LOCATION text box or by using the BROWSE button and selecting a folder. After creating a project, you will be presented with the screen shown in **Figure 2.2**.

This is the *Design Entry* GUI. Before inputting a schematic design, it is important for a user to become familiar with the icons, drop-down menus, and workspace. The icons along the right side of the screen are the action icons/buttons that allow for some of the most basic operations necessary for schematic design. These actions include cursor selection, wire/bus

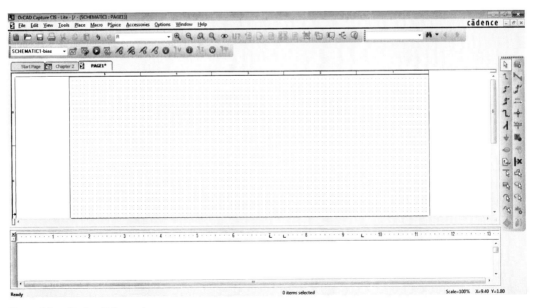

Figure 2.2 — *Design Entry* workspace.

placement, part placement, net alias naming, and ground placement and will be covered in this chapter (see **Figure 2.3**).

The buttons across the top are shortcut buttons for file management options, such as saving, opening, and creating new files, in addition to a ZOOM function. The ZOOM function can change the user's perspective of the schematic design: zooming in or out allows for a close-up view of a particular portion of the schematic or a wide view of the schematic as a whole. Other buttons along the top of the screen can be used to place measurement probes and create a simulation profile that defines the parameters of the simulation to be conducted on a given circuit (we will discuss all of these in the following examples).

The remaining central portion of the screen is the workspace, which is where the user can develop or draw a schematic design, placing and connecting components from the *PSpice* libraries within the software. Also, take notice of the text box in the lower right corner of the workspace. In this text box, you can provide a title and description for the schematic design. This can be especially helpful when performing analysis on complex projects. When working on a project with multiple pages and schematic designs, titles and descriptions can help you differentiate one design from another.

Before developing and testing a schematic design, you must ensure you have the desired components and libraries available in the *Design Entry* software. To do so, click the PLACE PART button on the toolbar at the right of the screen (shown highlighted in **Figure 2.4**). After clicking the PLACE PART button on the toolbar, the right side of the screen will open as shown in **Figure 2.5**.

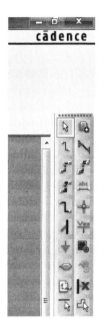

Figure 2.3 — DRAWING toolbar.

Figure 2.4 — PLACE PART button.

Figure 2.5 — ADD LIBRARY button.

The only default library available to a user is the DESIGN CACHE. The DESIGN CACHE will catalog and keep track of all of the existing components of a particular project, allowing for easy access to a component frequently used in a design. For example, if you are drawing multiple band-pass filters using inductors, capacitors and resistors in the *Design Entry* software, after you place one of each into the schematic, you will be able to quickly find these components in the Design Cache library for future use, rather than having to search through the libraries they originally came from. Since your new project has just been created, no components have been placed as yet, so your DESIGN CACHE will be empty. To proceed to the schematic design

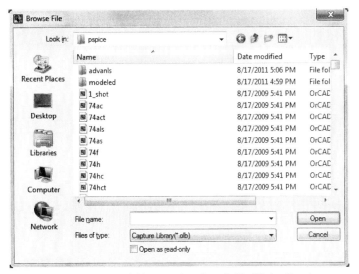

Figure 2.6 — BROWSE FILE folder contains all available *PSpice* libraries.

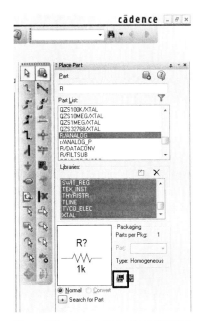

Figure 2.7 — Active component libraries.

process you must open the desired component libraries. To open a library, click the ADD LIBRARY button, highlighted in Figure 2.5. You then will be presented with the window shown in **Figure 2.6**.

In the *PSpice* folder, select the component libraries that you wish to add to the *Design Entry* software. For simulation purposes, you must be sure that you are adding only the libraries from the *PSpice* folder. Other libraries included with the Cadence software may include the same components utilized in your schematic design, but they can't be used for simulation purposes; they are merely schematic symbols used to generate schematic designs. After selecting the libraries that you wish to add, click the OPEN button. Now all the components from the selected libraries are available and can be placed into a schematic design (see **Figure 2.7**).

The text box titled LIBRARIES displays the libraries that have been selected from the *PSpice* folder. The box above that, titled PART LIST, displays all available components from the selected libraries. There are two ways you can select a part that is to be placed in a schematic design. First, by scrolling through the PART LIST with the up and down arrows, you may select a part for placement; once you have chosen a part, double-click the part for placement. Second, if you know the name of a component in the PART LIST, you can type it into the PLACE PART text box and press the ENTER key. After double-clicking on the part or typing in a component name and then pressing ENTER, the selected component is now attached to the mouse cursor. Move the mouse cursor over the drawing area and place the component by clicking the mouse on the destination of choice in the work space.

Referring again to Figure 2.7, all the component libraries have been selected, thus all their components are displayed on the PART LIST. To illustrate the SEARCH function, the character R has been entered into the PART text box. In the Cadence software, the resistor is cataloged in the *PSpice* library using this abbreviation. The resistor component is immediately highlighted on the PART LIST. At this point, the user can either press ENTER or double-click the highlighted component to attach it to the mouse cursor for placement. Notice the highlighted icon in Figure 2.7; this is the *AMS Simulator* icon. The presence of this icon while a component is highlighted in the PART LIST tells you that this component can be used for simulations and circuit analysis. If this icon is not present, it may indicate that a library has been added that has parts that are not compatible with the *AMS Simulator.* While it may be used to create a schematic drawing, the part cannot be used for analysis.

Drawing a Schematic Design

Let's begin with a dc circuit containing a battery (independent voltage source) and three resistors. The circuit diagram is shown in **Figure 2.8**.

This schematic design can be entered into the *Design Entry* software.

The two component types utilized in this circuit are the dc voltage source and the resistor. As we learned, the abbreviation R is used to catalog the resistor in the PART library. The dc voltage source is cataloged under the abbreviation of VDC. On the page, place one dc voltage source and three resistors.

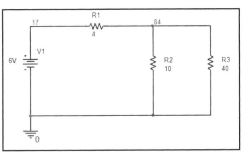

Figure 2.8 — Basic schematic design using one dc source and three resistors from *PSpice* libraries.

Once the components have been placed, they can be repositioned to allow for the configuration shown in Figure 2.8. Notice that the default placement of a resistor in the *Design Entry* software is horizontal, similar to the position of R1 in the schematic shown. In order to place the resistors vertically, similar to R2 and R3 in the parallel branches of the schematic, simply right-click the component and select ROTATE. Another way to rotate a component is to click on the part so that is highlighted and then press the R key on your keyboard. The default cursor in the *Design Entry* software is the SELECT tool, which enables the user to place and move parts. The SELECT tool is enabled in **Figure 2.9**.

Figure 2.9 — SELECT tool has been enabled in DRAWING toolbar.

To change the cursor to the WIRE tool, select the icon directly below the SELECT tool on the toolbar to the right of the screen. Once the four components are properly oriented, they can be connected with the WIRE tool to resemble the schematic diagram in the example shown in Figure 2.8. Click and hold the left mouse button while drawing lines to connect the component terminals (**Figure 2.10**), and you will see that components will be placed with a default value. In the case of the dc voltage source, the default voltage value is 0 V; in the case of the resistor, the default resistance value is 1 kΩ.

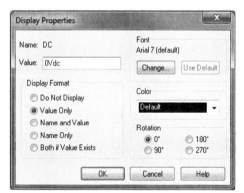

Figure 2.10 — Components have been placed into the DESIGN ENTRY window, but component values and net names have not yet been declared.

These values can be modified to fit the parameters and specifications of our schematic design in Figure 2.8. To modify a component value you may employ one of two methods. First, using the SELECT tool, hover the cursor over the line of text that declares a value next to the component. Double-click the line of text and you will be presented with a window titled DISPLAY PROPERTIES (see **Figure 2.11**). By double-clicking 0Vdc, the default voltage value, the DISPLAY PROPERTIES window is now active. To change the value of potential that is present in this dc voltage source, simply enter a new voltage value into the VALUE textbox and press the OK button. For this particular schematic design, the value of the dc voltage source happens to be 6 V. Using this same method, change the value of R1 from 1 kΩ to 4 Ω, R2 to 10 Ω, and R3 to 40 Ω.

There's another method that can be used to display and alter all component parameters. Use the SELECT tool to highlight a component and then double-click on it, which will open the PROPERTY

Figure 2.11 — DISPLAY PROPERTIES window for dc voltage source V1.

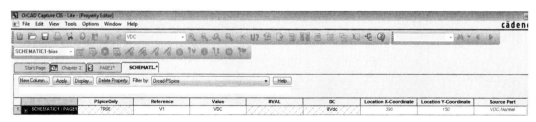

Figure 2.12 — PROPERTY EDITOR window.

EDITOR tab, as shown in **Figure 2.12**. With the PROPERTY EDITOR tab open, you can change more than one parameter at a time. After the parameters have been altered, click the APPLY button to save changes to the component. With the correct components, values, and wiring, the schematic begins to take shape (see **Figure 2.13**).

Before moving on to the simulation, you must make two more alterations to the schematic. To run any type of simulation using the *AMS Simulator*, the circuit needs a reference ground. Looking back to the original schematic, note that the circuit ground is attached to the negative terminal of the voltage source. To replicate this in the *Design Entry* software, look to the toolbar to the right of the screen and click the button with the ground schematic symbol (see highlighted portion of **Figure 2.14**). You will be presented with another part directory in a window titled PLACE GROUND (see **Figure 2.15**).

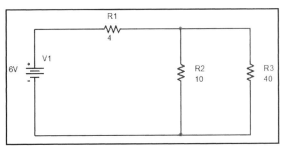

Figure 2.13 — Schematic design from Figure 2.10 has been modified to declare new component values.

The ground symbol is cataloged using the number 0. Place and connect this schematic symbol to the rest of the schematic design as seen in Figure 2.8. The schematic design is almost complete, but you can make one last alteration to your circuit by adding Net Aliases. Net Aliases are names chosen by the user to reference any node in a given circuit. The example schematic shown in Figure 2.8 names the node at the positive terminal of the voltage source 17, and the node between

Figure 2.14 — PLACE GROUND button.

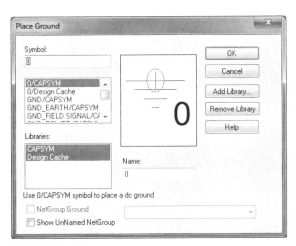

Figure 2.15 — PLACE GROUND window.

Figure 2.16 — NET ALIAS button.

Figure 2.17 — NET ALIAS window.

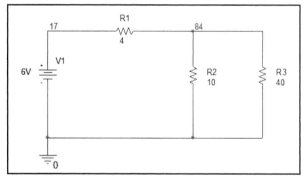

Figure 2.18 — Completed schematic design.

R1 and the parallel branches is named 84. To insert these node names into the *Design Entry* schematic design, click the NET ALIAS button located on the toolbar to the right of the screen (see **Figure 2.16**).

Next, open the PLACE NET ALIAS window. In the ALIAS text box, the user is allowed to enter any desired alphanumeric value (see **Figure 2.17**). Pressing OK will attach the user-defined node name to the cursor, and it can be placed on the schematic page at any node of the schematic design. After placing and connecting the ground symbol and Net Alias names to the schematic, it should resemble the schematic shown in **Figure 2.18**.

In Summation

This chapter has outlined some of the fundamental concepts of schematic design using *Cadence Allegro Design Entry* or *OrCAD Capture*. It is crucial that the user have a strong understanding and a good familiarization with the schematic workspace to take full advantage of the Cadence software package. The next chapter provides a further explanation of the interconnections between the components of the schematic drawing seen in this chapter and describes these components and interconnections in the form of the circuit's netlist.

Chapter 3

Netlist Element Lines

In the previous chapter we introduced the Cadence *Allegro Design Entry* (or *OrCAD Capture*) software and demonstrated the basics of how to create a schematic design. In the *Design Entry* workspace we created a circuit consisting of three resistors and a dc voltage source; this circuit has been recreated as shown in **Figure 3.1**.

Generating a Netlist from a Schematic Design

Before we move into the discussion of simulation types and simulation profiles that can be set up within the Cadence software, it is important that you have a rudimentary understanding of how the software will interpret this schematic design. To expand upon this topic, we will introduce the concept of the circuit's *netlist*.

A netlist is a text file that provides a complete list of all of the components used in a schematic design. Using a specified notation, this list also describes all the different *nets*, or points of connection, between each of these components. Therefore, every circuit you create will have an associated netlist. We will begin by outlining the process involved in generating a netlist based on an existing schematic design, and then making this netlist visible to the user. After the netlist is generated, we'll see how to interpret the notation used in each element line of the netlist.

In this first example, we see that the circuit in Figure 3.1 has been drawn into a new project titled

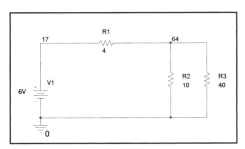

Figure 3.1 — Three-resistor circuit.

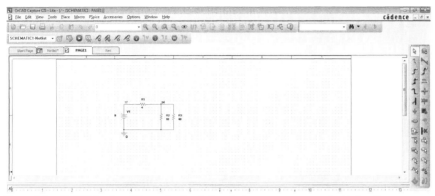

Figure 3.2 —*OrCAD Capture/Allegro Design Entry* window, SCHEMATIC project tab active.

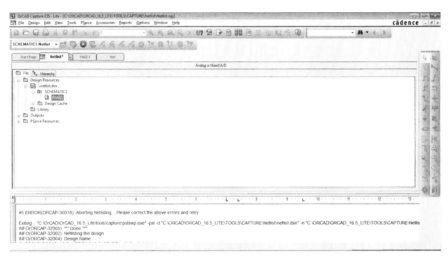

Figure 3.3 — *OrCAD Capture/Allegro Design Entry* window, PROJECT HIERARCHICAL tab.

"Netlist" (see **Figure 3.2**). After saving the circuit, click the project tab (labeled NETLIST) to display the project files (see **Figure 3.3**).

The design shown in Figures 3.1 and 3.2 is located on Page 1 of the project's schematic. Click the PAGE 1 icon to highlight this page, as shown in Figure 3.3. From within the toolbar that spans the top of the user interface, click the CREATE NETLIST button (available in the TOOLS menu)) to open the CREATE NETLIST window (see **Figure 3.4**).

Figure 3.4 — CREATE NETLIST window.

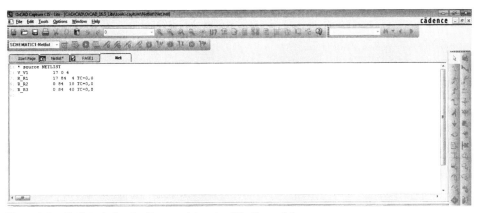

Figure 3.5 — Netlist window for three-resistor circuit in Figure 3.1.

Select the PSPICE tab in the CREATE NETLIST window. Under this tab, use the NETLIST FILE: textbox to enter a filename and destination to save the netlist file. For the purposes of this example, the netlist file has simply been titled *Net.net*. After clicking OK, the netlist file is created and will open as a new tab in the Cadence software (see **Figure 3.5**).

Interpreting the Netlist

Figure 3.5 presents the user with the netlist associated with the schematic from Page 1 of the schematic design. **Figure 3.6** is a close-up of the data displayed in Figure 3.5 and legibly displays the five lines that comprise the newly generated netlist.

The first line reads * source NETLIST and informs the user of the source schematic of the data displayed in the lines below. In this case, the source is the "Netlist" project created earlier. The next four lines of the netlist are referred to as element lines. These element lines reference the individual components of the schematic, outline their nodal connections, list their values, and can even allow the user to designate other modifications to the underlying parameters of each part. The second line reads V_V1 17 0 6. The first letter of this element line, V, designates that this element line indicates a voltage source. This is followed by an underscore, and then the name of the voltage source component as it is listed in the schematic design. Referring to the schematic design, the name of the dc voltage source is V1 and is visible in this element line.

Recall that when we created the schematic design for this three-resistor circuit, Net Alias names were provided at different nodes of the circuit. An alias name of "17" was designated at the node where the positive terminal of the dc source connects to the first terminal of R1, and an alias name of "84" was designated where the second terminal of R1 connects to the R2 and R3 resistive branches. With this in mind, let's look at the second half of this

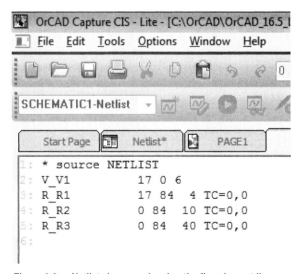

Figure 3.6 — Netlist close-up showing the five element lines.

element line, which reads 17 0 6. The first number, 17, is representative of the nodal connection of the positive terminal of the voltage source. The second number, 0, is representative of the nodal connection of the negative terminal of the voltage source. Even though the user did not actively create a net alias of "0" at the negative terminal, this is the default net name at the point of the circuit that has been designated as circuit ground. The third numerical value is the magnitude of voltage. In the case of this design, the dc source is set to 6 V.

The next three element lines reference the three resistors and delineate their interconnections in the schematic design. The third line in the netlist reads R_R1 17 84 4 TC=0,0. As was seen with the voltage source, the first letter of this element line designates the type of component that is being referenced. In this case, R means that this component is a resistor. This is followed by an underscore and then the name of the resistive component, R1, as listed in the schematic design. The three numerical values that follow the component name, 17 84 4, are net alias names and component values. The first number, 17, is net alias "17" and signifies the connection of the first terminal of resistor R1. Looking back to the voltage source element line, note that the positive terminal of the voltage source connects to net "17" as well; therefore, we see that the positive terminal of voltage source V1 connects to the first terminal of R1. The next number in the element line, 84, indicates the connection of the second terminal of R1 in the schematic design to net alias "84." The third value, 4, tells the user that component R1 is a 4 Ω resistor. The last portion of the element line reads TC=0,0. TC stands for temperature coefficient and will delegate certain physical changes in resistance with respect to change in temperature in increments of 1 kelvin. The default values for both temperature coefficient values are zero, so any simulation results yielded from this netlist would mimic ideal conditions.

The next two element lines map out the connections of the last two resistors in the circuit, R2 and R3. Looking at these two element lines, it is evident that these resistors are in parallel. The first terminal of R2 and R3 connects to circuit ground, and the second terminal of each resistor connects to net alias "84", which also happens to be the second terminal of R1. These element lines also describe the values of resistance for R2 as 10 Ω, and for R3 as 40 Ω. As was seen in the R1 element line, the TCs have not been changed from their default values of 0,0 for either R2 or R3.

Generating a Netlist Without Net Aliases

When creating a schematic design, it is not necessary to create net alias names for every node in the circuit. A netlist can be generated on a schematic with no designated net aliases. To illustrate this point, the net aliases "17" and "84" have been removed from the circuit shown in Figure 3.1 and the

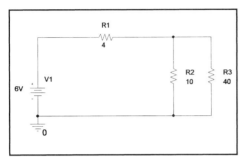

Figure 3.7 — Schematic of the three-resistor circuit in Figure 3.1 with net aliases removed.

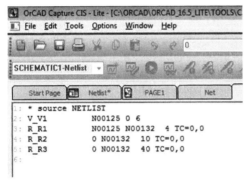

Figure 3.8 — Netlist for the schematic in Figure 3.7 with arbitrary net alias names assigned by the software to unnamed nodes.

schematic has been saved in the "Netlist" project (see **Figure 3.7**). The schematic layout remains the same, the net aliases have been removed, and a netlist has been generated (see **Figure 3.8**).

Upon generating a netlist or carrying out a simulation of a schematic design, the Cadence software will assign arbitrary net names to any unnamed node. Looking at the netlist from Figure 3.8, we see we are dealing with the same schematic design, components, component names, and values, but instead of listing net names "17" and "84", as seen in Figure 3.6, the user is provided with random net names, "N00125" and "N00132", respectively. After creating the netlist, the schematic design in the *Design Entry* window will not change, regardless of the fact that net names have technically been assigned. In effort to synchronize the schematic design with the newly created netlist, net names "N00125" and "N00132" have been entered manually into the schematic design (see **Figure 3.9**). Now it is easy to see the parallels between the original schematic with included net names and the second schematic with net aliases removed; in both situations the netlist is derived using the same basic format.

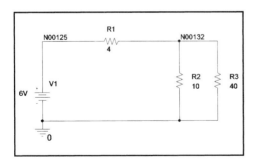

Figure 3.9 — Schematic design from Figure 3.7 changed to incorporate arbitrary net alias names.

RLC Netlist

To take this process one step further, a basic series RLC (consisting of a resistor, an inductor and a capacitor) schematic design has been created and net aliases provided (see **Figure 3.10**).

The first two examples gave the reader insight into the layout of the element lines used for both a voltage source and a resistor. With the introduction of a capacitor and inductor into the schematic design, the same general layout will be seen with the new element lines, but there will be a slight difference in the nomenclature used to describe these components. The netlist has been generated and can be seen in **Figure 3.11**.

The inductor element line reads L_L 2 3 10uh and follows the same format as the voltage source element line. The L indicates that this element line is that of an inductor. This is followed by an underscore, then the component name used in the schematic. The first two numerical values are the nodal connections of the two terminals of the inductor, and the third number represents the component's value, in this case, 10 μH.

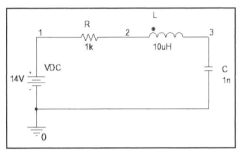

Figure 3.10 — Series RLC circuit.

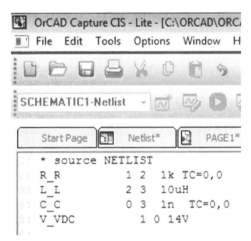

Figure 3.11 — Netlist for the RLC circuit from Figure 3.10.

The capacitor element line reads C_C 0 3 1n TC=0,0 and closely follows the format of the voltage resistor element line. The letter C indicates that this element line is that of a capacitor. This is followed by an underscore, then the component name used in the schematic. The first two numerical values are the nodal connections of the two terminals of the capacitor, and the third number represents the component's value, in this case, 1 nF (1n in the netlist). Just like with the resistor, the temperature coefficient (TC) of the capacitor can be modified so that the capacitor more closely mimics the behavior of a real capacitor as opposed to an ideal one. This TC models a change in capacitance with respect to a change in temperature in increments of 1°C.

In Summation

At first glance, it may not seem important to dissect the netlist as we did in this chapter. After all, it is easy enough to use the design entry features of the *OrCAD* software to draw a circuit to be simulated, but a thorough understanding of the netlist will certainly help you with other aspects of design and troubleshooting. A netlist provides an excellent representation of all the interconnections within a design as well as the fundamental parameters of these components. The netlist concepts discussed in this chapter will be referenced later in this text.

Chapter 4

Basic Simulation Types

We now have a complete schematic design drawn in the Cadence *Allegro Design Entry* or *OrCAD Capture* schematic capture software, complete with node names, designated component values, and ground. At this point, the user is free to begin any desired simulations or circuit analysis of the schematic design. Later in this chapter, readers will be introduced to some of the fundamental circuit simulation and analysis techniques that can be utilized in Cadence *Allegro Design Entry* and *Allegro AMS Simulator* (or *OrCAD Capture* and *PSpice A/D*). To explore these different simulation types, sample simulations will be performed on the simple three-resistor circuit design already constructed in Chapter 2 and shown in Figure 2.18, in addition to three other schematic designs.

Bias Point

The first simulation to be conducted on the three-resistor circuit is that of a Small Signal Bias Solution or Bias Point Calculation. This type of simulation will perform voltage calculations from each node of the circuit with respect to ground, current calculations within the circuit, and the

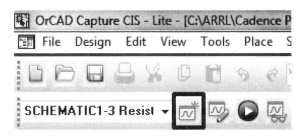

Figure 4.1 — NEW SIMULATION profile button.

calculated power dissipated in each component of the circuit. Before the circuit can be simulated, the user must set up a simulation profile.

To create a new simulation profile click the SHORTCUT button highlighted in **Figure 4.1**. Clicking the NEW SIMULATION button presents the window shown in **Figure 4.2**, which prompts you to enter a title for the simulation you wish to run. Any alphanumeric combination can be entered into the title box. In the NEW SIMULATION window, you see that the name "3 Resistor Circuit" has been chosen for this example, and it merely serves as a description of the schematic design. After choosing a name for the simulation profile, click the CREATE button, which causes a window titled SIMULATION SETTINGS to open (see **Figure 4.3**). Note that "Simulation Settings" in the title of this window is followed by the name chosen in the new simulation window, in the case of this example that is "3 Resistor Circuit". This is a helpful feature offered by the Cadence software, because it allows the user to be mindful of the simulation parameters that are being altered. In a case where analyses are being performed on a more complex design, the user can have more than one simulation profile in a project.

Now let's turn to the first simulation to be demonstrated, the Small Signal Bias Solution. In the active SIMULATION SETTINGS window, select the ANALYSIS tab. Under that tab is a drop-down menu labeled ANALYSIS TYPE. Open this drop-down menu and select BIAS POINT to apply the changes that you made to the simulation profile. Now the simulation parameters are in place to run the Small Signal Bias Solution. After clicking the OK button, you will be returned to the DESIGN ENTRY window. To run the simulation, refer to the toolbar that spans the top of the screen and click the button highlighted in **Figure 4.4**.

Notice that prior to creating a simulation profile, this button is grayed out and cannot be selected. Only after creating the simulation profile as described in the previous paragraph will this button be enabled. After clicking on the SIMULATION button, wait a few moments for the simulation process to complete. You will

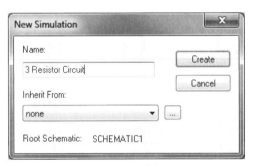

Figure 4.2 — NEW SIMULATION window.

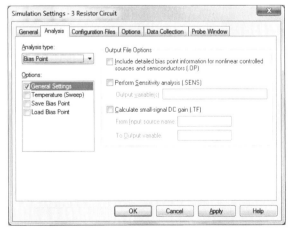

Figure 4.3 — SIMULATION SETTINGS window, Bias Point.

Figure 4.4 — RUN PSPICE button.

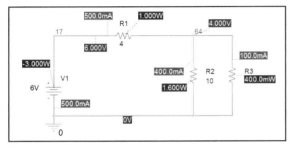

Figure 4.5 — Bias Point results displaying voltage, current and power.

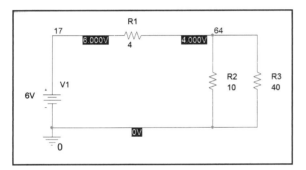

Figure 4.6 — Bias Point results displaying only voltages.

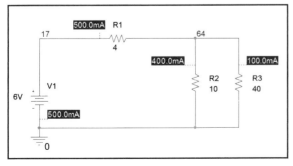

Figure 4.7 — Bias Point results displaying only currents.

know that the simulation is complete when the progress bar displayed on the screen reaches 100%. Referring to Figure 4.4, note the three shortcut buttons to the right of the SIMULATION button, labeled V, I and W. These buttons toggle the labels for the calculated values of voltage (V), current (I) and power (W) gathered from the simulation results. These values are displayed as labels on the initial schematic design. All three measurements have been toggled on and are visible in the simulation output shown in **Figure 4.5**.

In **Figure 4.6** only the voltage measurement button has been enabled; therefore only values of voltage are visible on the schematic drawing. These voltage values are the amount of potential present from the labeled point with respect to ground. In **Figure 4.7** only the current measurement button has been enabled; therefore only values of current are visible on the schematic drawing. These current measurements are the amount of current present in the labeled path. In **Figure 4.8** only the power measurement button has been enabled; therefore only values of power are visible on the schematic drawing. These values label the individual components of the circuit with the power dissipated within them. When you compare the simulation results seen in Figures 4.6, 4.7, and 4.8 to the results seen in Figure 4.5, you will notice that the user can either display or hide whatever values are desired at a given time. Similar to moving a component or component label during the schematic design process, measurement labels can be moved to any destination on the

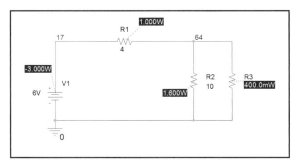

Figure 4.8 — Bias Point results displaying only wattages.

schematic page just by clicking and dragging. After the simulation has been completed, if you wish to modify your design to obtain a different wattage, voltage or current, you must edit the desired component values using the method discussed in Chapter 2; you then simply run the simulation again.

DC Sweep

The next type of simulation covered in this chapter is the DC Sweep. Using the methods discussed in earlier chapters, create either a new project or simply add a new schematic page using the option in the file menu, and then draw the circuit shown in **Figure 4.9**.

This is a three-resistor circuit with two dc voltage sources. As we saw in the previous simulation, the user is able to perform a Small Signal Bias Solution to determine the voltage, current, and power in various points of the circuit. The DC Sweep simulation allows you to vary a specific component value and display an output plot of current, voltage or power based on the specified or incremental changes in component value. After creating the circuit shown in Figure 4.9, you must create a new simulation profile. This can be done in the same manner as it was in the previous simulation. For this example, the simulation has been titled "DC SWEEP 2V 3R". After naming the simulation, you are presented with the SIMULATION SETTINGS window (see **Figure 4.10**).

In the SIMULATION SETTINGS window, change the analysis type to DC Sweep using the drop-down menu to display the simulation parameters shown in Figure 4.10. For this simulation we will be sweeping the voltage value for the dc voltage source labeled VRIGHT. To sweep this component value, select VOLTAGE SOURCE in the SWEEP VARIABLE portion of the window. In the NAME text box to the right enter the component name that you wish to sweep exactly as it appears in the schematic design. For this simulation, we will sweep VRIGHT from 0 V to 10 V in 0.5 V incremental steps. To define

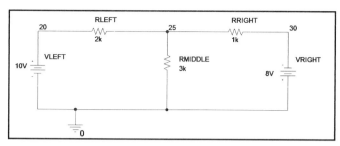

Figure 4.9 — Two-voltage source circuit.

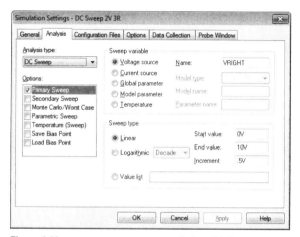

Figure 4.10 — SIMULATION SETTINGS window, DC Sweep.

these parameters within the simulation profile, enter a start value of "0V", an end value of "10V", and an increment of ".5V" into their respective text boxes. Since the sweep is across only a small range of voltage, select a LINEAR sweep type. A LOGARITHMIC sweep can be selected to accommodate a wide range between start or end values whether you are sweeping across voltage, current or resistance.

Notice the text box at the bottom of the window labeled VALUE LIST; if that is selected, it enables you to enter user-defined component values, and the simulation will sweep across these values only while generating an output. After entering the desired DC Sweep parameters, apply the changes to the simulation profile and run the simulation. After the simulation is complete, the *Allegro AMS Simulator* will open, displaying a blank page where the output plot will be displayed (see **Figure 4.11**).

The simulation has performed a nodal analysis on the circuit as the value of VRIGHT changes from 0 V to 10 V. Using the *AMS Simulator* we will note the impact of the change in voltage across the resistor RMIDDLE

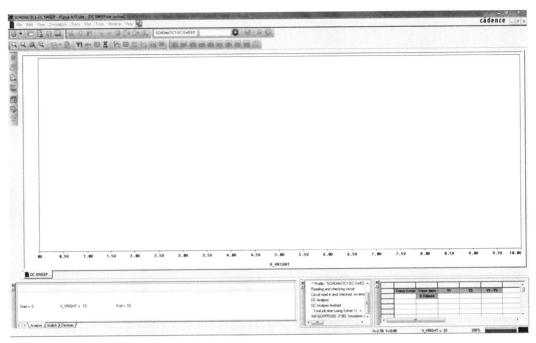

Figure 4.11 — *PSpice A/D/Allegro AMS Simulator* window, also referred to as PROBE.

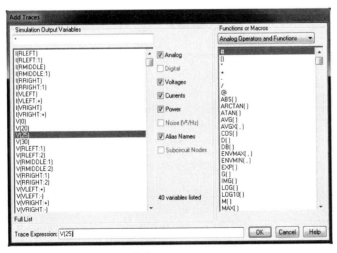

Figure 4.12 — ADD TRACES window.

as the sweep takes place. To observe this change, select ADD TRACES from the TRACE menu. This will open the window shown in **Figure 4.12**.

The left column displays a list of measurements of voltage, current and power that can be taken at any given node in the circuit. For example, if there is a trace expression V(R1:1) listed in the column, it simply means that the trace to be plotted will show a voltage measurement from terminal 1 of component R1 with respect to ground. Also take notice that if you provided a net alias to any node of the circuit in the schematic design, it will be displayed in this list. In the case of this schematic design the node between RLEFT and

RRIGHT has been given a net alias of "25". In order to display a plot of the voltage across RMIDDLE, either find V(25) on the TRACE EXPRESSION list or enter it into the text box as shown. Press ENTER to plot the trace. The ADD TRACES window will close and the plot will appear in the *AMS Simulator* window (see **Figure 4.13**).

The X axis labeled V_VRIGHT displays the swept voltage range defined in the SIMULATION SETTING window as "0V" to "10V". The Y axis displays the range of calculated voltage across RMIDDLE. At any point, the user can double-click any open space surrounding the X or Y axis to the AXIS SETTINGS menu (see **Figure 4.14**).

The AXIS SETTINGS menu allows the user to customize the *AMS Simulator* workspace by selecting grid line types using the options under the X GRID and Y GRID tabs or axis labels under the X AXIS and Y AXIS tabs.

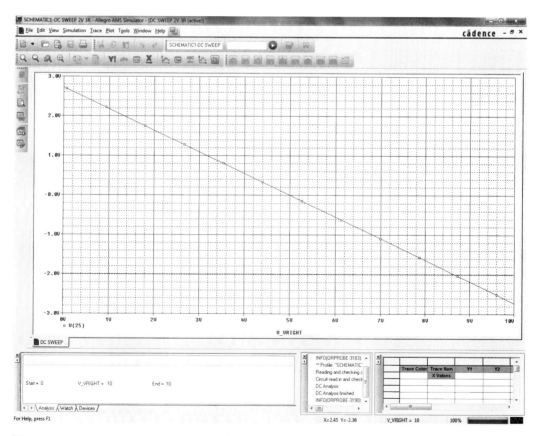

Figure 4.13 — Plot of voltage measured from node 25 with respect to ground.

Using this window, the Y axis has been given the title "Voltage Across RMIDDLE" (see **Figure 4.15**).

You can now begin to interpret and analyze these simulation results. One way to do so employs the CURSOR function of the *AMS Simulator*. To enable the cursors, click the CURSOR button highlighted in Figure 4.15. Note the presence of two cursors on the voltage trace. The cursors can be moved to any data point on the trace visible on the plot. The left mouse button will move one cursor, and the right mouse button will move the other. Figure 4.15 illustrates the operation of the cursors on a trace and demonstrates how to mark a data point. Two random data points have been selected using each cursor. To the right of the button that enabled the cursors, you will find a bank of buttons that allow you to position cursors on various positions of the trace. The button to the far right will mark data points in ordered pair format at the position of each cursor. In the bottom right corner of this window, note the presence of a

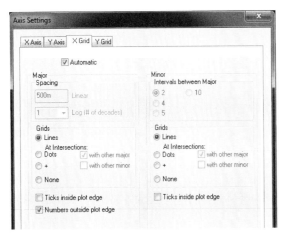

Figure 4.14 — AXIS SETTINGS window.

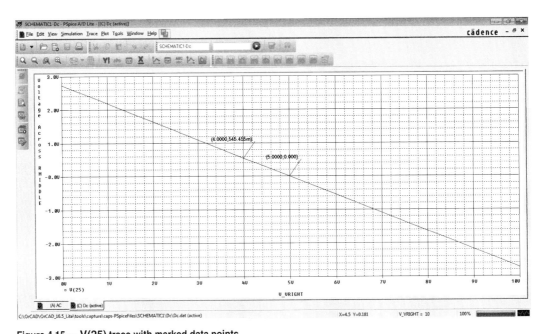

Figure 4.15 — V(25) trace with marked data points.

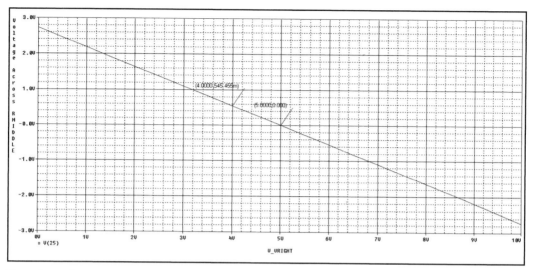

Figure 4.16 — Plot copied to clipboard.

small spreadsheet window. This spreadsheet window will identify the X and Y values based on individual cursor positions, allowing you to view exact values according to simulation results without having to mark a multitude of data points. However, for the purpose of this example, two data points have been marked based on cursor position. At the instance when component VRIGHT is equal to 4.1770 V, the voltage across resistor RMIDDLE is 448.914 mV, and when VRIGHT is equal to 5.1770 V, the voltage across resistor RMIDDLE is 96.514 mV.

Based on the screenshot of Figure 4.15, you may find it hard to decipher marked data values or read a plot with multiple traces. To make this process easier you may refer to the AXIS SETTINGS window mentioned earlier and implement changes to the gridlines to make the plot more legible. If you wish to copy the output plot of the *AMS Simulator* to another document, simply use the WINDOW menu and select COPY TO CLIPBOARD; this will open an active window that lets you select the color scheme for the copied plot image. To demonstrate these changes and how they can be applied, the generated plot using these features (a black and white image with minor grid lines removed from the X and Y axis) is shown in **Figure 4.16**. Note that the layout of the trace, axis, and labels is identical, but the plot is much easier to read. Note also that the X and Y values of the two marked data points remain intact as well.

AC Sweep

The next analysis to be discussed is the AC Sweep. This simulation allows the user to provide a constant value of ac voltage to a circuit while varying input frequency across a defined range. **Figure 4.17** is a schematic design of a simple RLC tank circuit.

This circuit design uses components that have not been used or seen in previous examples. These components, as listed in the *PSpice* component library, are the ac voltage source VSIN, capacitor C, and inductor L. In creating the schematic, you can define these component values using the same process described in the beginning of the chapter. Using the ac voltage source VSIN, you have the option of applying an offset voltage and setting amplitude, frequency and ac voltage. You're given the option to apply an ac voltage value and amplitude because the ac voltage value is used as a constant value for the AC Sweep simulation and because the values of amplitude, frequency and offset voltage are used for other simulation types. One of these is Transient Analysis, which will be discussed at the end of this chapter.

Note also the presence of the resistor component R2. This is not a typical design for an RLC resonant tank circuit, but the Cadence software will require the presence of this resistor as it sees the inductor in the same series path as an ideal component with no internal resistor. Component R2 merely simulates the internal resistance of a real inductor. Running a simulation without this series resistor will result in an error, since the Cadence software will see a branch with a 0 Ω internal resistance. Once the circuit is complete, create a new simulation profile just as was done in the previous examples. Again, you will open the SIMULATION SETTINGS window after naming the simulation (see **Figure 4.18**).

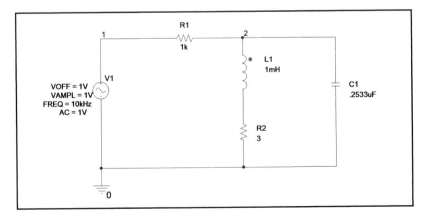

Figure 4.17 — RLC tank schematic.

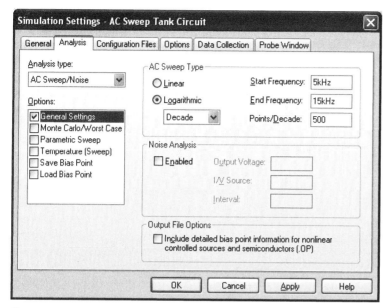

Figure 4.18 — SIMULATION SETTINGS window, AC Sweep.

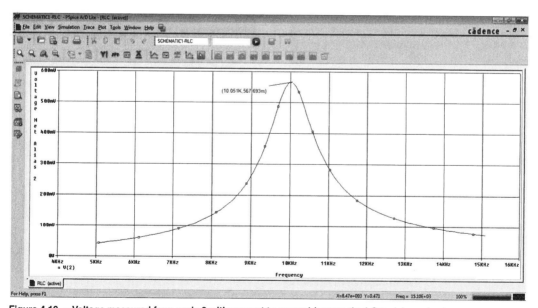

Figure 4.19 — Voltage measured from node 2 with respect to ground (across the LC tank branches).

For the purpose of this example, the simulation has been titled "AC Sweep Tank Circuit," as indicated in the title bar of the SIMULATION SETTINGS window. Using the ANALYSIS TYPE drop-down menu, select AC SWEEP/NOISE. Enter a start frequency, where the *AMS Simulator* will begin plotting measurements, and an end frequency, where the *AMS Simulator* will stop. The text box labeled POINTS/DECADE will allow you to define the resolution of the plotted trace. This value represents the number of data points per decade across the frequency spectrum; the greater the number of data points, the higher the resolution and definition of the plotted trace. Points per decade can be changed to points per octave depending on how you wish to view your output plot. Lastly, since we are sweeping across a wide frequency range, the logarithmic sweep type has been selected. After applying these changes, the circuit is ready for simulation.

Run the simulation and plot the voltage measured from net alias "2" using the TRACE EXPRESSION text box in the ADD TRACES window. Again, this trace expression, V(2), can be found either in the list to the left of the ADD TRACES window or entered into the text box. This trace represents the output voltage of the resonant tank circuit as the frequency changes from 5 kHz to 15 kHz. The X axis of this plot represents frequency and the Y axis represents the measured voltage from net alias "2". To provide a more user-friendly, legible output plot, the grid lines have been altered and the Y axis has been titled "Voltage Net Alias 2" (see **Figure 4.19**).

In the previous example, the CURSOR tool was briefly mentioned; we'll enable that now. The bank of buttons to the right of the CURSOR ENABLE button in the top toolbar accesses the cursor placement tools. The button to the immediate right of the CURSOR ENABLE button will drop the cursor at the maximum voltage value of this parabolic curve. In Figure 4.19, the maximum value of voltage has been marked using the button to the far right of the toolbar. This maximum value occurs at the resonant frequency of this circuit. Since inductive reactance increases as frequency increases and capacitive reactance decreases as frequency increases, there will only be one frequency where these two reactances will be equal.

Again, the peak of this curve indicates the resonant frequency and has been marked with an ordered pair. The X value of the ordered pair at the peak of the curve is measured at 10,022 kHz. This tells us that resonance occurs at this frequency. The measured voltage at this resonant frequency, the Y value of the ordered pair, is 568.809 mV. This plot is once again generated in grayscale using the COPY TO CLIPBOARD function and can be seen in **Figure 4.20**.

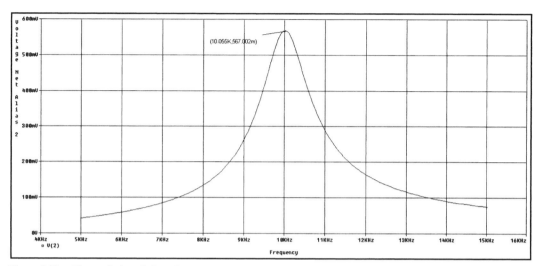

Figure 4.20 — Plot copied to clipboard.

Transient Analysis

In the final example of this chapter, an RC circuit will be connected to a pulse voltage source, and an analysis of circuit voltage versus time, called a Transient Analysis, will be performed. The pulse voltage source has not been seen in prior examples. The component is listed in the *PSpice* library as VPULSE. The result of the Transient Analysis will appear in the output file in tabular form and as a graph of voltage versus time. The circuit is shown in **Figure 4.21**.

The single pulse input source goes from 0 V to 1 V after 10 µs of

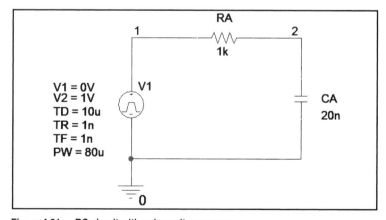

Figure 4.21 — RC circuit with pulse voltage source.

delay (relative to the beginning of the Transient Analysis), with rise and fall times of 1 ns, and has a duration of 80 μs. As in the previous examples, you must create a new simulation profile. Start by naming your simulation. For the purpose of this example, the simulation has been titled, "RC Pulse Transient," as indicated in the title bar of the SIMULATION SETTINGS window (see **Figure 4.22**).

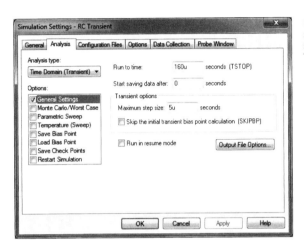

Figure 4.22 —
SIMULATION
SETTINGS window,
Transient Analysis.

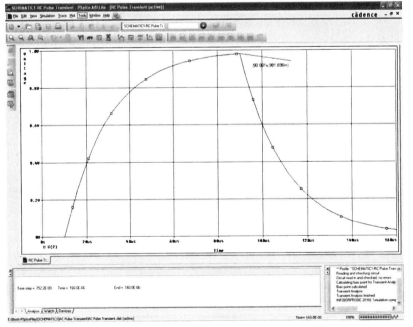

Figure 4.23 — Voltage
measured from node
2 with respect to
ground.

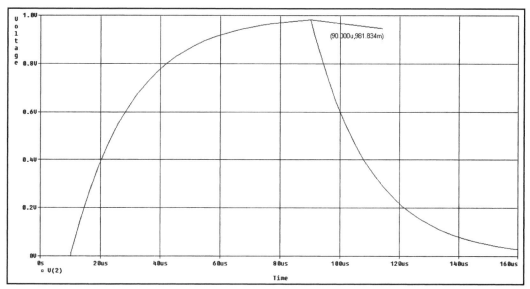

(90.000u,981.834m)

Figure 4.24 — Plot copied to clipboard.

In the SIMULATION SETTINGS window, select the TIME DOMAIN analysis type using the drop-down menu. To accommodate the pulse width defined in the schematic design, a total run time has been set to 160 µs. Data will be calculated and plotted starting at time 0 s. The step size will define the resolution of the trace displayed on the plot, and a smaller step size will yield a higher resolution as there will be smaller incremental steps between plotted data points. The step size chosen for this simulation is 5 µs. After applying these changes, the circuit can be simulated.

Figure 4.23 displays the measured voltage from net alias "2" from the RC circuit. This plot demonstrates the change in voltage as time progresses from 0 s to 160 µs. The Y axis has been labeled "Voltage," minor gridlines have been removed to make the plot easier to read, and the maximum voltage has been marked. The ordered pair reads (90.001u, 981.835m). This indicates that the maximum voltage, 981.835 mV, occurs at 90.001 µs. Lastly, the plot has been pasted in grayscale with these format changes in **Figure 4.24**.

In Summation

These four basic simulation types are the crux of this text, and it is of the utmost importance that you develop a level of comfort with these before reviewing the examples in later chapters. Being able to carry out each of these simulation types can tell you a lot about the behavior of a particular circuit.

Chapter 5

Netlist Control Lines

By now you should be getting accustomed to operating the *PSpice* user interface. You have been presented with an introduction to a circuit's netlist and are familiar with the most basic simulation procedures that the *PSpice* software has to offer. Now that we have covered these basic simulation techniques, we can delve further into the topic of the netlist.

The Complete Netlist — Bias Point Simulation

As discussed in Chapter 3, you can develop a schematic layout in a text format using what we have referred to as element lines. In addition to these element lines, control lines can be added to a netlist to describe a simulation type and to list its parameters. In this chapter, we will review each of the simulations covered in the previous chapter and view the control lines that were inadvertently added to the netlist during the simulation process.

Our first circuit (see **Figure 5.1**) is the simple three-resistor circuit used to demonstrate the Small Signal Bias Solution. In the last chapter all the circuit currents, voltage drops and power dissipations were made available in the *Allegro Design Entry* or *OrCAD Capture* software. Now we will view the simulation results attached to the netlist in the Cadence *Allegro AMS Simulator* or *PSpice A/D* window. The circuit shown in Figure 5.1 had been drawn into the *Design Entry* software, saved as "3 Resistor Circuit" and simulated using a Small Signal Bias Solution simulation profile titled "3 RESISTOR." After running the simulation, you will see the simulation results just as you did in the previous

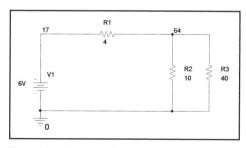

Figure 5.1 — Three-resistor circuit schematic.

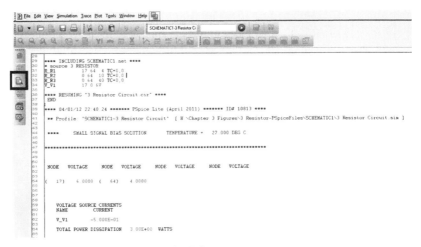

Figure 5.2 — Simulation output file, Bias Point.

chapter, but take note of the blank *AMS Simulator* window that opened after the simulation was completed. Bring the *AMS Simulator* window to the front and click the button highlighted in **Figure 5.2** on the toolbar to the left side of the screen. This button is labeled VIEW SIMULATION OUTPUT FILE and displays a text representation of the simulation result. After clicking this button, the blank *AMS Simulator* window will change and you will be presented with a window that closely resembles that shown in Figure 5.2.

The top portion of the screen should look familiar. It is the same netlist that was tested in the netlist chapter. This netlist is followed by a simple command, .END. This signifies the end of the netlist and will be seen at the end of every circuit netlist that is simulated in the software. This particular simulation type has no control line. With no control line present, *PSpice* will carry out a Small Signal Bias Solution by default and will consider the presence of a dc voltage source and perform the associated dc bias calculations at each node of the circuit. In the instance of a circuit that is purely ac, you will see a Small Signal Bias Solution in this window, except that all of the dc values will be listed with a magnitude of 0.

Just to clarify and help you clean up the layout of this netlist, note that any lines of the netlist that begin with * or a series of multiple * characters (such as *****) are insignificant to actions of the netlist. The * symbol denotes a comment line and can provide the user with helpful notes regarding the construction of the circuit or parameters of the simulation. The comment lines displayed above them in this figure are a direct result of simulating the circuit using the *Design Entry* schematic capture software and are generated at the same time that the *PSpice* software creates the netlist. In these comment lines, we have the date of the simulation, the destination of the source file, a title heading for the Small Signal Bias Solution results, and the default environment temperature of the simulation.

At the bottom of the screen, each node voltage is listed, followed by the total circuit current and the total power dissipated in the circuit. These are the same values that were seen in the previous chapter when the simulation results were viewed in the *Design Entry* window.

The Complete Netlist — DC Sweep Simulation

Next we will examine the control line used in the netlist for the DC Sweep simulation type. We will follow the same succession as we did with the previous chapter, so the second circuit will be used to illustrate this concept. The circuit shown in **Figure 5.3** was drawn in the *Design Entry* software, saved as "DC Sweep 2V 3R", and simulated using a DC Sweep simulation profile titled "DC SWEEP". As in the last chapter, the voltage source VRIGHT will be swept from 0 V to 10 V in increments of 0.5 V. These values were entered into the SIMULATION SETTINGS window prior to the simulation of this circuit.

As usual, you are presented with a blank *AMS Simulator* window. As was done with the first circuit, click on the VIEW OUTPUT FILE button at the left of the screen to view the output file (see **Figure 5.4**). At the bottom portion of the screen, the element lines of the netlist are visible. The control line is seen toward the top portion of the screen and reads as follows:

.DC LIN V_VRIGHT 0V 10V .5V

The first component of the control line, .DC, designates that the simulation will be a DC Sweep; LIN means that this will be a linear sweep; next it is declared that the component to be swept across is the voltage source, VRIGHT, referenced in the netlist as V_VRIGHT. The last three numerical

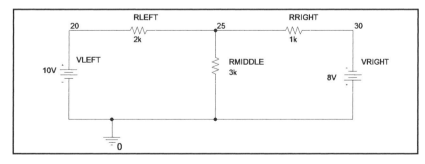

Figure 5.3 — Two-voltage-source circuit.

Figure 5.4 — Simulation output file, DC Sweep.

values are the parameters that were entered into the SIMULATION SETTINGS window, and they state that the voltage will be swept from a starting value of 0 V dc to an ending value of 10 V in 0.5 V increments.

The next line, .PROBE, writes the simulation analysis results to an output data file. The line following this, .INC, declares the source schematic file. At the very bottom of the window we see the .END command signifying the end of the netlist.

The Complete Netlist — AC Sweep Simulation

Next we will inspect the control line used in the netlist for the AC Sweep simulation type, referring to the third circuit from the last chapter to demonstrate the usage of this control line. The circuit shown in **Figure 5.5** was drawn in the *Design Entry* software, saved as "AC Sweep Tank", and simulated using an AC Sweep simulation profile titled, "AC SWEEP TANK CIRCUIT".

After running the simulation and opening the simulation output file, you will be presented with the window shown in **Figure 5.6**. Similar to the output file of the DC Sweep, the control line is at the top of the window, the netlist is in the middle, and the Small Signal Bias Solution is at the bottom. The circuit shown in Figure 5.5 utilizes an ac voltage source, V1. This ac source outputs a 1 V_{pk-pk} sinusoidal waveform with a +1 V dc offset. As

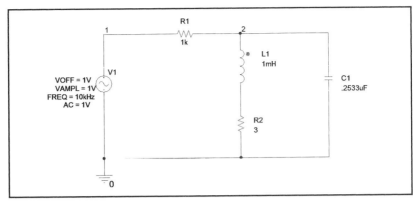

Figure 5.5 — RLC tank circuit.

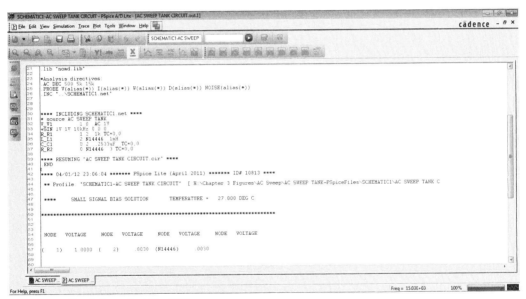

Figure 5.6 — Simulation output file, AC Sweep.

previously stated, the *PSpice* software will always perform a Small Signal Bias Solution. The presence of this offset voltage will contribute dc nodal voltages that will be seen in the Small Signal Bias Solution.

The construction of the netlist is relatively straightforward, and we now understand that the node voltages seen at the bottom of the window are a direct result of the dc offset. This brings us to a discussion of the AC Sweep control line. The AC Sweep control line reads as follows:

.AC DEC 500 5K 15K

The beginning of this control line, .AC, tells us that an AC Sweep will be performed, and DEC declares that the X axis of this frequency response plot will be divided into decades, as opposed to octaves, across the logarithmic scale. These prompts are followed by three numerical values that further define the parameters of the AC Sweep simulation. These are the same values that you entered into the simulation settings window in the *Design Entry* software. The numerical value 500 states that there will be 500 data points per decade, and the last two values, 5K and 15K, set the range of the frequency sweep that occurs during the simulation.

The Complete Netlist — Transient Analysis Simulation

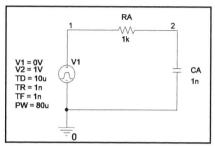

Figure 5.7 — RC circuit with pulse voltage source.

Now we move on to the fourth circuit of the previous chapter, which simulated a simple pulse voltage across an RC circuit. The circuit shown in **Figure 5.7** was drawn into the *Design Entry* software, saved as "RC Transient" and simulated using a Transient Analysis simulation profile titled "PULSE TRANSIENT".

Upon opening the simulation output file shown in **Figure 5.8**, you will observe a layout similar to what was seen with the previous simulations. The control line is near the top of the window, the netlist is near the middle,

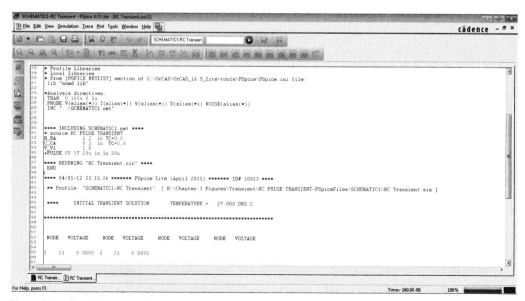

Figure 5.8 — Simulation output file, Transient Analysis.

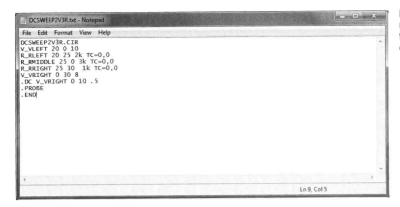

Figure 5.9 — Complete netlist file for DC Sweep of two-voltage-source circuit, created in *Notepad*.

and the Small Signal Bias Solution is at the bottom. We saw the impact that an offset voltage could have on the results of a Small Signal Bias Solution in the previous example. In this example, there is no dc potential; regardless, the software carries out the Small Signal Bias Solution and labels the value of voltage at each node as 0 V.

Looking to the top of the window, we see the control line for the Transient Analysis simulation profile, which reads as follows:

.TRAN 0 160u 0 5u

The command .TRAN means that the simulation will be a Transient Analysis. The next two numerical values define the range of time for the simulation. In accordance with the example, the simulation will run from 0 to 160 μs. The next number, 0, means that simulation data will be saved starting at the time of 0 s. The last number, 5u, is the step size, which as discussed in the previous chapter, helps define the resolution of the output plot.

Simulating a Circuit Using a Netlist File

After reviewing all the examples put forth in this chapter, you may be wondering why it is necessary to be able to interpret netlist data if the *PSpice* software can create netlist data for you during the simulation process. With a solid understanding of the concepts of different element lines and control lines, you do not need to draw a schematic into the *Design Entry* software and use the schematic capture function of the *PSpice* software to generate and simulate the netlist. You can begin by creating the netlist in a text document and simulate the circuit using just the *Allegro AMS Simulator*. This procedure will be demonstrated below using the dc circuit from the second example (Figure 5.3).

In **Figure 5.9**, a text file using the *Notepad* program has been created and saved as "DCSWEEP2V3R". The top line of the netlist that we will create will be the name of the circuit file that will be simulated. You may choose any name or reference you like, but it must be followed by the suffix ".CIR" (DCSWEEP2V3R.CIR). Next we have the element lines that describe the interconnections and values of the components that make up the schematic design. In an effort to keep this example as simple as possible, the names referencing the individual components of the Figure 5.3 circuit have been kept the same and match the element lines shown in the simulation output file in Figure 5.4. Next we have the DC Sweep control line followed by our probe command and end command. *Notepad* saves files as text format (.txt extension). After saving the *Notepad* document, using your desktop browser, go to the destination of the text file and change the extension from ".txt" to ".cir."

When creating a netlist, it is important to use a simple text editor such as Microsoft *Notepad* or *WordPad*. If a user tries to create a netlist in a more complex word processing program such as Microsoft *Word*, Corel *Word Perfect*, or Apple *Pages*, any simulation will likely fail due to the presence of hidden characters saved in the document for page formatting purposes.

Next, open the *Allegro AMS Simulator* and select OPEN in the FILE drop-down menu. Select the destination of the netlist text document that you have just created. In **Figure 5.10**, the newly created text file is visible in the

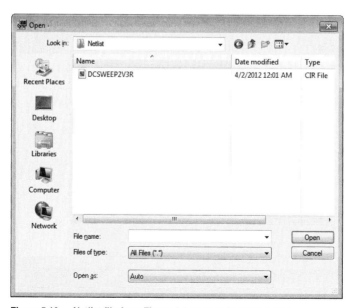

Figure 5.10 — Netlist file from Figure 5.9 as seen in OPEN window.

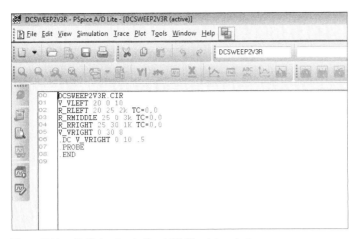

Figure 5.11 — Netlist active in the *AMS Simulator* window.

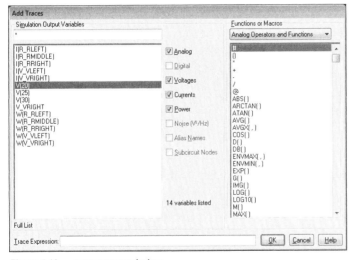

Figure 5.12 — ADD TRACES window.

OPEN window. Note that since the file extension has changed, the icon now resembles that of a schematic design file; note also that under the file type column, this document is listed as a "CIR File". Open this file to create an active window that resembles **Figure 5.11.**

The netlist that was just created using the text editor is now visible in the *AMS Simulator* window. At the top of the window, press the green button labeled RUN. This will simulate the circuit file outlined in the netlist as "DCSWEEP2V3R".

After the simulation is complete, you will be presented with a blank output plot as in the previous chapter. Now it is easy enough to resume using the software as if you had just simulated a schematic using the *Design Entry* software. Open the ADD TRACES menu (see **Figure 5.12**). Since our text netlist kept the same component names and node names, we see all of the available traces that we saw when we ran the DC Sweep in the simulation chapter.

Figure 5.13 is a screen capture of our netlist simulation with a voltage trace measured from node 25 with respect to ground while VRIGHT is sweeping. This plot shown below yields results that are identical to the plot of this circuit from the previous chapter generated from a schematic capture.

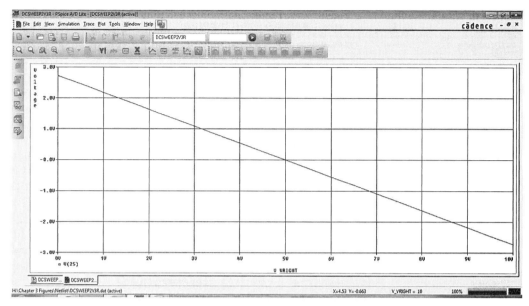

Figure 5.13 — Probe graph developed using the netlist file.

In Summation

This concludes the overview of the netlist. Between the material represented in this section and Chapter 3, a user can begin to understand how the layout of a schematic design and corresponding simulation profile can be written in text form. An understanding of these concepts can only better help you carry out routine simulation procedures using the *OrCAD* software and the *PSpice* user interface.

Chapter 6

The *PSpice* Probe Tool

The two main goals of the previous chapter were to familiarize the user with the Cadence user interface and present four basic examples incorporating different circuits and simulation types. In this chapter, we will explore some of the capabilities of the Cadence *Allegro AMS Simulator* or *PSpice A/D* utilizing the PROBE tool in the Cadence *Allegro Design Entry* or *OrCAD Capture* software.

Adding Measurement Markers to a Schematic Design

To begin, the circuit shown in **Figure 6.1** is constructed within the *Design Entry* software. This circuit uses the same component values for the

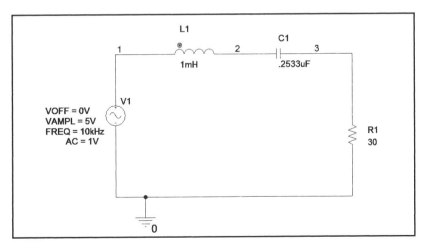

Figure 6.1 — RLC tank circuit schematic.

capacitor and inductor as the RLC tank example from Chapter 3; however, this circuit maintains a series configuration.

A Transient Analysis will be performed on the circuit, plotting a sinusoidal output in the *AMS Simulator*. To achieve the desired output so that we have more than two completed cycles of the output waveforms, a runtime of 300 µs is declared. The plot will begin at the time of 0 s, and to provide an adequate resolution to the output plots, a step size of 300 ns has been selected. The simulation settings are shown in **Figure 6.2**.

After running the simulation, the *Allegro AMS Simulator* will automatically open. After the simulation is complete, you can bring the *Design Entry* window to the front and display the original schematic design. In **Figure 6.3**, PROBE tools have been highlighted in the toolbar that spans the top of the screen. These probes designate the output traces that will be plotted in the *AMS Simulator* window.

The first probe to be placed on the schematic design will be a single-voltage measurement probe. Select the PROBE tool to the far left in the bank

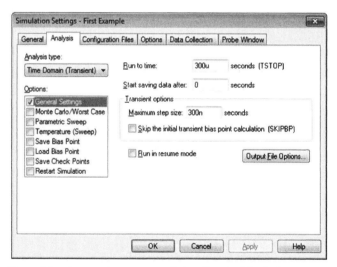

Figure 6.2 — SIMULATION SETTINGS window, Transient Analysis.

Figure 6.3 — Measurement markers.

of probes. Hovering the cursor over this PROBE icon will display its name, VOLTAGE/LEVEL MARKER. This voltage probe is placed on the node in the circuit labeled as net alias "1" (see **Figure 6.4**). When a single voltage probe is placed into a schematic design in this manner, a voltage measurement will be taken from this point with respect to ground. The placement of this voltage probe will plot the source voltage.

Once the voltage probe has been placed, the trace will appear in the *AMS Simulator* window. Bring the *AMS Simulator* window to the front and observe the presence of the new source voltage trace, which will be named "V(1)". In **Figure 6.5**, a $1V_{pk}$ waveform is now visible.

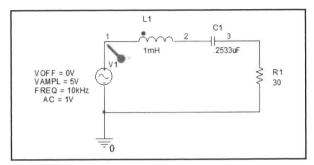

Figure 6.4 — Schematic diagram with voltage marker placed at node 1.

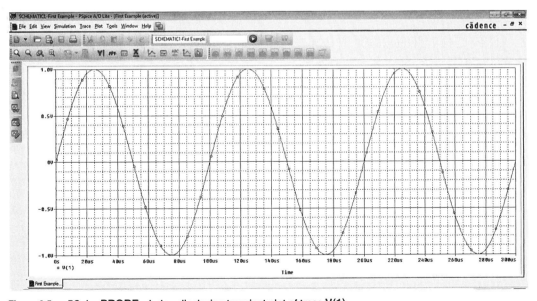

Figure 6.5 — *PSpice* PROBE window displaying transient plot of trace V(1).

After moving back to the *Design Entry* window, hover the cursor over the next probe. This probe tool can be referred to as the VOLTAGE DIFFERENTIAL MARKER(S). Select this probe, which allows you to place two probes into the schematic diagram. The output trace that will be plotted as a result of this probe placement will be a measure of the voltage present between the two markers. Note that there is both a positive and negative marker. A voltage trace will be plotted in the *AMS Simulator* window based on a voltage measurement taken from the node placement of the positive marker with respect to the node placement of negative marker. One could compare the placement of these voltage probes to the placement of voltmeter probes. This set of differential markers has been placed around the inductor, L1, as shown in **Figure 6.6**.

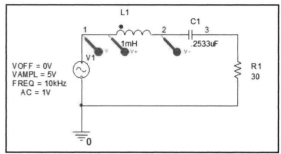

Figure 6.6 — **Voltage differential marker placed across component L1.**

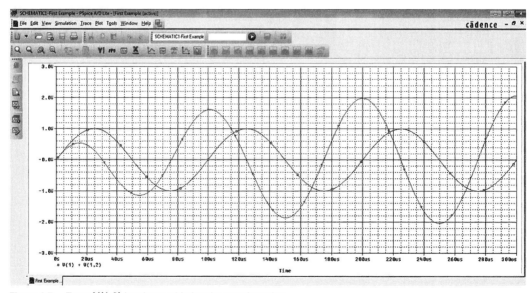

Figure 6.7 — **Trace V(1,2) has been added to the transient plot.**

With both the voltage and differential probes in place, two traces are now visible in the *AMS Simulator* window. The original 1 V_{pk} waveform remains, and the new trace that has been added to the plot is that of the voltage across the inductor. Refer to the bottom left corner of the screen and note the name of the trace as V(1,2) (see **Figure 6.7**).

One last set of differential markers has been placed into the schematic design to add one last trace to the plot. This set of markers will lie on nodes 2 and 3, respectively, and provide a trace that plots the voltage across the capacitor (see **Figure 6.8**).

The two previous traces remain and this third and final trace has been

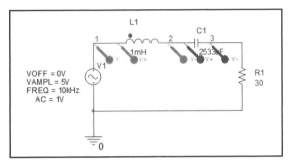

Figure 6.8 — Differential marker placed across component **C1**.

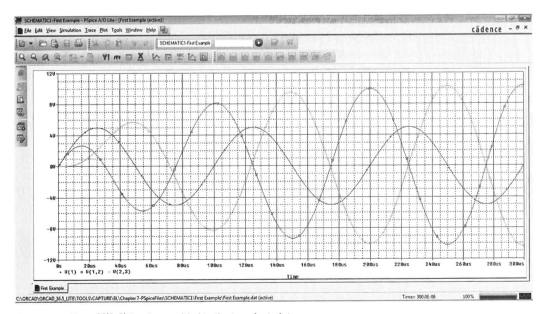

Figure 6.9 — Trace V(2,3) has been added to the transient plot.

added to the plot. This trace plots the voltage across the capacitor. Refer to the bottom left corner of the screen and note the name of the trace as V(2,3) (see **Figure 6.9**).

Adding Labels

The *AMS Simulator* is user-friendly in the sense that it lets you manipulate the plot in certain ways that allow the information and data to be presented more neatly. The plot shown in Figure 6.9 could be considered cluttered and illegible. At first glance, it may be difficult to decipher which plot is which. We will begin by providing labels to each of the traces on the plot. In the, PLOT drop-down menu, find the TEXT option under LABEL. The user will be presented with the window shown in **Figure 6.10**.

Enter an alphanumeric combination into the text box to generate a label

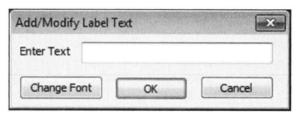

Figure 6.10 — ADD/MODIFY LABEL TEXT window.

that can be placed onto the plot. After clicking OK or pressing enter, the designated label will be attached to the mouse cursor and can be placed anywhere the user desires. In **Figure 6.11**, three labels have been added to the plot. Each trace has been labeled with a description in addition to the trace name.

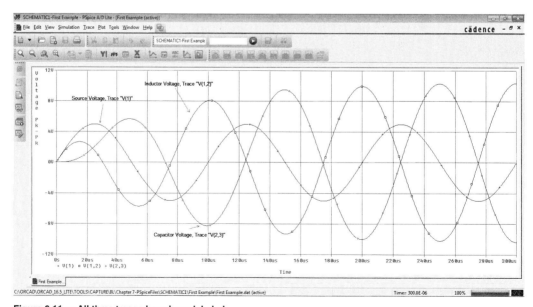

Figure 6.11 — All three traces have been labeled.

The three labels read as follows:
"Source Voltage, Trace V(1)"
"Inductor Voltage, Trace V(1,2)"
"Capacitor Voltage, Trace V(2,3)"

Notice that arrows have been inserted to designate which trace corresponds to which label. To insert an arrow, open the PLOT drop-down menu and find ARROW under LABEL (see **Figure 6.12**). Using the mouse

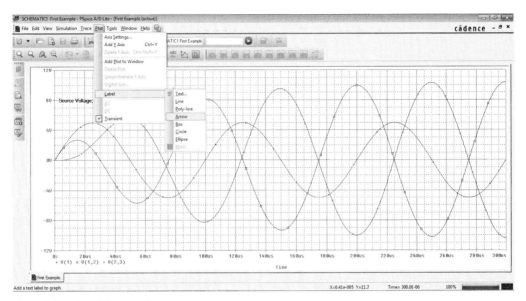

Figure 6.12 — Insert Arrow option.

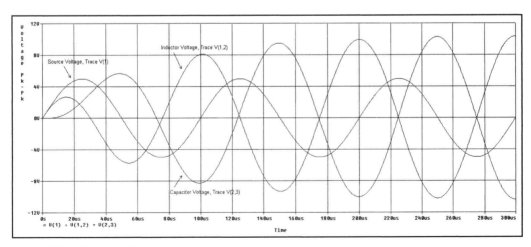

Figure 6.13 — Plot copied to clipboard in grayscale.

cursor, select the desired start and end points on the plot to draw the arrow in place.

Now the axes are labeled and the plot has been output in grayscale. This finished product can be seen in **Figure 6.13**.

Adding a Second Plot to the *AMS Simulator* Window

This first example uses a transient response simulation to plot the output voltage across the capacitor and inductor with a plot of the source voltage visible simultaneously. The frequency of the source voltage is 10 kHz. This also happens to be the resonant frequency of the RLC circuit, so when plotting the output, you might find it beneficial to isolate the inductor and capacitor voltages from the source voltage using separate plots; this lets you clearly observe the phase relationship between the inductor and capacitor when the RLC circuit is at resonance. As previously stated, the *AMS Simulator* can allow the user to display the output data and plots in a variety of different ways.

In the next example, the same circuit and traces will be employed in generating output data in the *AMS Simulator*, but this time the traces will be contained within two plots in the same window.

To accomplish this, it is not necessary to run the simulation again; rather you will be manipulating the data present after the initial run of the Transient Analysis. The first step is to delete the traces from the *AMS*

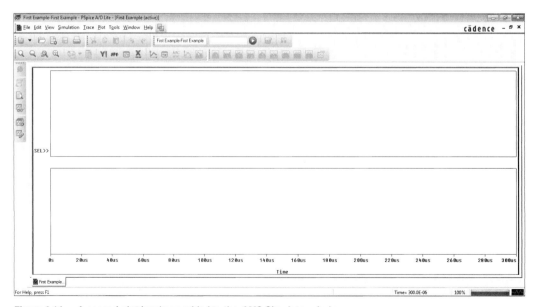

Figure 6.14 — A second plot has been added to the *AMS Simulator* window.

Simulator window. This can be done in one of two ways. One method is to click the individual trace names at the bottom left of the screen and press the DELETE key; the other method is to select the option DELETE ALL TRACES under the TRACE drop-down menu. After doing so, the user will see a blank plot with an undefined Y axis and an X axis ranging from 0 s to 300 μs. By selecting the option ADD PLOT TO WINDOW under the PLOT drop-down menu, a second plot will appear in the workspace, as shown in **Figure 6.14,** with each plot taking up approximately half the active window.

The axis labels and grid lines can be adjusted accordingly for each independent plot. This means that the plot in the top half of Figure 6.14 can have a unique gridline arrangement and Y axis title while the bottom plot can have its own configuration. The method in which these parameters can be altered remains the same. Again, to change an axis title or modify the gridlines, simply double-click on either Y axis to open the AXIS SETTINGS menu.

Now with two plots to work with, the traces can be reinserted. First we will input the capacitor and inductor traces into the bottom plot. In Figure 6.14, take note of the SEL>> indicator, next to the Y axis of the top plot. Any traces added in the *AMS Simulator* will be to the plot associated with the SEL>> indicator. In the case of Figure 6.14, traces would be added to the top plot. Select the bottom plot by clicking the blank region to the left of the lower Y axis, then add traces V(1,2) and V(2,3), as in **Figure 6.15**. In this figure, gridlines and labels have been put in place. After moving the

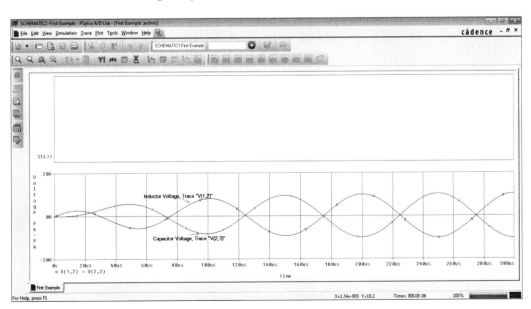

Figure 6.15 — Traces V(1,2) and V(2,3) have been added to the bottom plot.

SEL>> indicator to the top plot, the source voltage trace, V(1), can be plotted.

In **Figure 6.16**, all traces have been plotted and labeled. Each Y axis is labeled and both plots have the appropriate gridlines.

Finally, the grayscale plot has been generated in **Figure 6.17**. With the capacitor and inductor voltages isolated in the bottom plot, it is very easy to see the phase relationship between these two waveforms. As previously

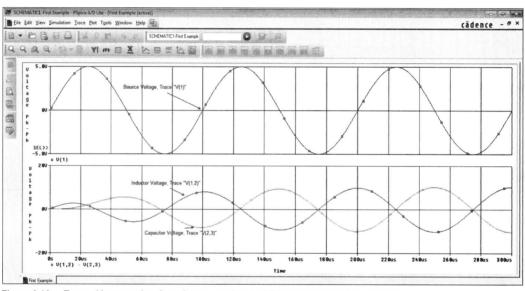

Figure 6.16 — Top and bottom plots have been completed adding all traces and axis labels.

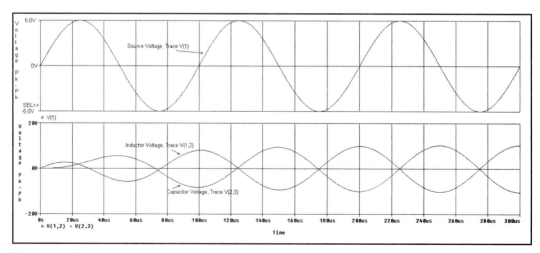

Figure 6.17 — Plot copied to clipboard in grayscale.

stated, the RLC circuit is running at its approximate resonant frequency of 10 kHz. At resonance, the reactances for both the capacitor and inductor are identical, so the voltage across each component will be of the same magnitude. It is important to remember that although the magnitudes of the V(1,2) (inductor) trace and the V(2,3) (capacitor) trace are similar, these waveforms are 180° out of phase with each other.

This will lead to a cancellation between the capacitor and inductor waveforms, leaving a purely resistive circuit. This phase shift can easily be seen in the bottom plot of Figure 6.17. With this modified layout, one could argue that it is much easier to interpret output data when looking at Figure 6.17 rather than Figure 6.13, where all three traces were on top of each other.

Adding Traces Manually Using Analog Operators and Functions

The marker tools that have been used thus far and that have been placed in the *Design Entry* software have allowed the user to measure units of voltage. While these probe tools may seem rather simple, used in conjunction with ANALOG OPERATORS AND FUNCTIONS in the ADD TRACE window, the user can plot almost any characteristic or function imaginable. Using this second circuit, a new RLC resonant one, we will create another multi-plot window showcasing both the impedance and phase angle of a tank circuit with respect to change in frequency. The circuit to be constructed in the *Design Entry* software is shown in **Figure 6.18**. This circuit is identical to the resonant tank circuit from the previous chapter.

Figure 6.18 has been drawn in a slightly different manner. Note the

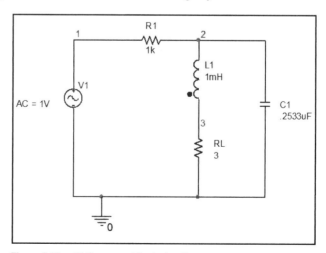

Figure 6.18 — RLC resonant tank circuit.

schematic symbol of our ac source voltage, V1. The VSIN component will have associated parameters of amplitude, frequency and offset voltage as well. For this particular example, we will be performing an AC Sweep. The only relevant parameter for this simulation type and example is the one shown, AC=1V. A value of 0 V has been entered for a dc offset, and miscellaneous values are input for the remaining parameters of amplitude and frequency since they will not impact the results of the AC Sweep. To neatly present this schematic, the only value that has been labeled is the aforementioned AC=1V. After the schematic is drawn, the simulation parameters are entered for the AC Sweep (see **Figure 6.19**).

After entering the simulation parameters, run the simulation. As in the previous example, the *AMS Simulator* will open with a blank plot. Begin by adding a second plot to the window. Select the top plot and open the ADD TRACE menu. In the text box at the bottom of the window, manually add the trace name, "V(2)/I(V1)". Taking the voltage across the LC parallel branches, V(2), and dividing by the total current in the circuit, I(V1), will calculate the impedance of the tank circuit. As the frequency sweep is occurring, the inductive reactance and capacitive reactance will change. These constant changes in inductive reactance and capacitance will

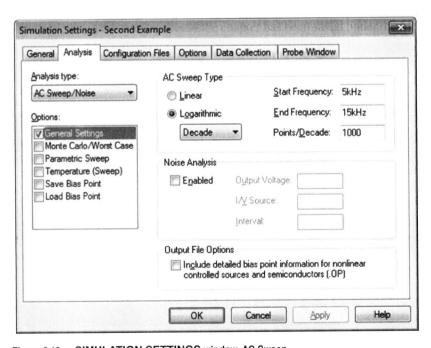

Figure 6.19 — SIMULATION SETTINGS window, AC Sweep.

im-pact the impedance of the circuit and will be reflected in the output plot. The resonant frequency of this non-ideal tank circuit is approximately 10 kHz. At resonance, the inductive and capacitive branches will be seen as having an impedance of $(Z) = \infty$, with an equivalent parallel resistance of 1.318 kΩ. This is shown in the top plot of **Figure 6.20**.

Next, the bottom plot will be added to the *AMS Simulator* window.

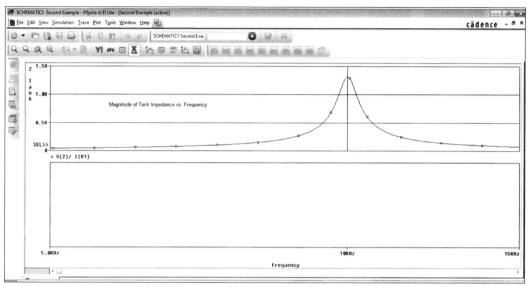

Figure 6.20 — Plot of tank impedance with respect to frequency.

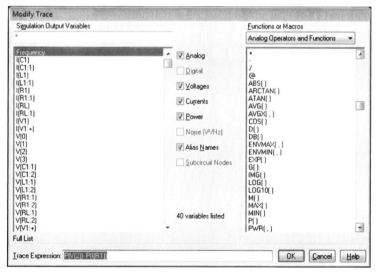

Figure 6.21 — Adding a trace using the phase operator.

This will be a plot of the phase angle of the tank circuit with respect to frequency. Select the bottom plot and open the ADD TRACE window. The right column of this window is titled ANALOG OPERATORS AND FUNCTIONS (see **Figure 6.21**).

Scroll down and find the operator P(). This operator will plot the phase angle of the trace entered within the parentheses. Click on this term or manually enter the characters "P()" into the trace text box at the bottom. Between the parentheses, enter the trace name "V(2)". From this first term, "P(V(2))", subtract "P(I(R1))". The full trace name becomes "P(V(2))–P(I(R1))" and is nothing more than the phase of the circuit current subtracted from the phase of the tank circuit voltage. The plot of this trace will illustrate the change in phase shift as the frequency sweep occurs. This final plot is shown in **Figure 6.22.**

The final plot has been generated in grayscale in **Figure 6.23**. With the two plots adjacent to each other, it seems easier to interpret the relationship between these two plots. In the top plot, the labeled peak magnitude of the tank impedance occurs at the resonant frequency of the RLC circuit (10 kHz). At resonance, there will be no phase shift between the circuit current and the voltage of the tank circuit, so we can note that at this same frequency in the lower plot the phase angle measures 0°. At this point we know that the circuit is purely resistive. At any frequency below resonance, the RLC circuit is capacitive; therefore, it is indicative in the lower plot

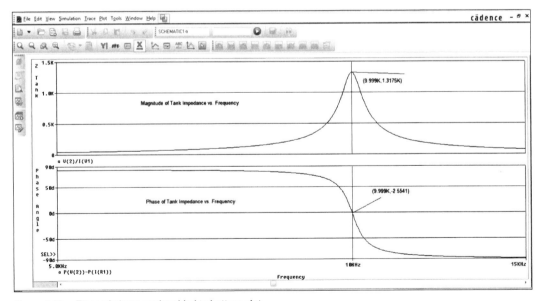

Figure 6.22 — Trace of phase angle added to bottom plot.

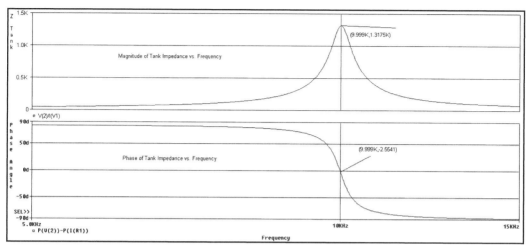

Figure 6.23 — Plot copied to clipboard in grayscale.

that the phase angle begins to climb in the positive direction. Likewise, at frequencies above resonance, the trace in the lower plot begins to climb increasingly in the negative direction when the circuit is inductive.

Transient Analysis Results in the *AMS Simulator* Window

The next example will be that of a simple series RC circuit with a dc voltage applied. After constructing a schematic, a Transient Analysis will be completed to plot voltage across the individual components of a circuit as the capacitor charges. Again, we will employ two separate plots in the same window, one of which will be used for the voltages previously alluded to and the other for power and current. To begin, the schematic design was constructed in the *Design Entry* software as shown in **Figure 6.24**. After drawing the schematic design, double-click on the capacitor to open the parameters for the component. Find the parameter IC (Initial Charge) and input a value of "0". This assures that the simulation will begin with an uncharged capacitor.

Next a simulation profile is created entering the simulation settings shown in **Figure 6.25**. The profile is set to run a Transient Analysis with a runtime of 50 ms beginning at time, t = 0s.

After running the simulation, the *AMS Simulator*

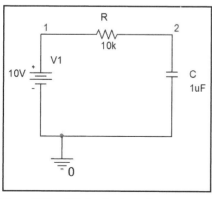

Figure 6.24 — RC circuit schematic.

will open. Just as was done in the previous examples, add a second plot to the window. Select the top plot and, using either the VOLTAGE DIFFERENTIAL MARKER probe tools in the *Design Entry* window or by entering the traces manually, plot the resistor voltage and the capacitor voltage. The bottom plot will contain two traces. One will be the current in the series path and the other will be a measure of power across the capacitor. To reiterate, either method can be used to plot each trace. The probe tools can be placed in the *Design Entry* software, or the trace names can be input in the ADD TRACE window. Now a total of four traces can be seen within the two plots (see **Figure 6.26**).

In Figure 6.26, the top plot shows the capacitor voltage, trace V(2), and the resistor voltage, trace V(1,2). In the bottom plot, the circuit current, trace I(C) is accompanied by the trace of the power dissipated in the capacitor, W(C). Notice, in the bottom plot, there is no unit labeled on the Y axis, merely a prefix for milli. The axis has been simply titled "Magnitude" because there are two traces based on different fundamental units. Therefore, it is understood that when interpreting the current trace, measurements will be taken in the unit of milliamperes, and when interpreting the power trace, the unit will be expressed in terms of milliwatts. Finally, the plot is generated in grayscale in **Figure 6.27**.

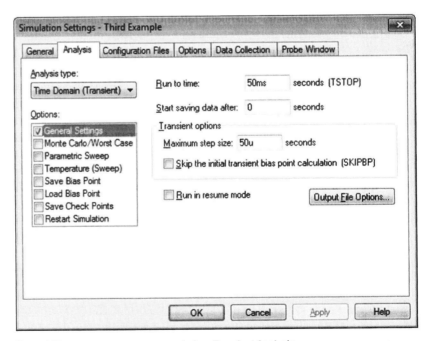

Figure 6.25 — SIMULATION SETTINGS window, Transient Analysis.

The bottom plot is rather self-explanatory. It contains two traces, one of circuit current and one of power delivered to the capacitor with respect to time. The top plot allows the user to see the relationship between capacitor voltage and resistor voltage. As the capacitor continues to charge and store potential, the voltage drop across the resistor will continue to decrease.

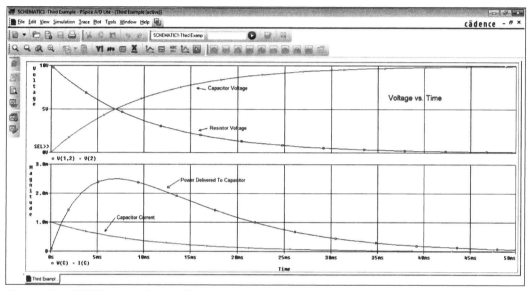

Figure 6.26 — *AMS Simulator* window with graphs of voltage, current and power.

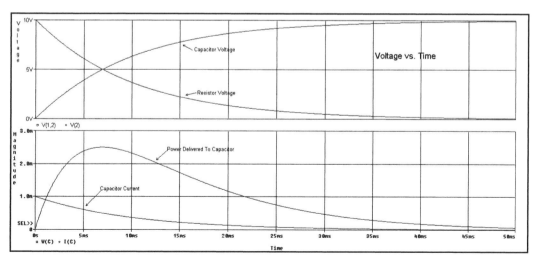

Figure 6.27 — Plot copied to clipboard in grayscale.

Transient Analysis and Fast Fourier Transform

Our fourth example is another Transient Analysis that plots the cycles of a square wave. The circuit for this is a single voltage source connected to a load resistor (see **Figure 6.28**). The type of voltage source used is the component labeled VPULSE. It is crucial to modify the parameters of the pulse source in order for the output voltage to resemble that of a square waveform. To do so, a quick rise time and fall time of 1 ns have been designated. The pulse width is 0.5 ms, which is half of the period of the waveform, and the amplitude of this waveform has been defined as 1 V, the difference on V1 and V2.

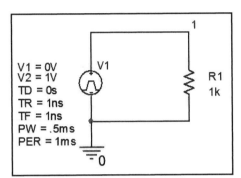

Figure 6.28 —
Schematic of pulse voltage source across single-load resistor.

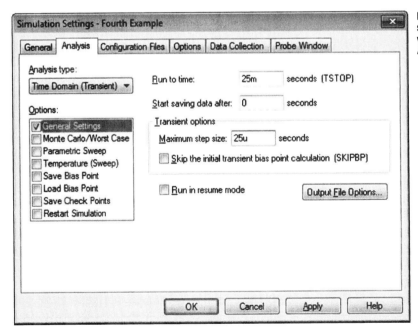

Figure 6.29 —
SIMULATION SETTINGS window, Transient Analysis.

Figure 6.29 shows the SIMULATION SETTINGS window. The simulation profile is set to perform a Transient Analysis with a runtime of 25 ms starting at time, t = 0s. The output plot will display 25 cycles.

Because of the simplicity of this output waveform, the next screenshot is the generated grayscale plot (see **Figure 6.30**). The plot consists of the plotted load voltage with respect to time.

The fact that this last example utilizes a square waveform allows us to explore one of the other features of the *AMS Simulator*, the Fourier Analysis Function. A Fourier Analysis determines the harmonic content of a complex waveform. In the SIMULATOR window, the plot of a 1 kHz square waveform is visible. A square wave is composed of a sine wave accompanied with multiple odd harmonics of that fundamental frequency. By carrying out a Fourier Analysis of the plotted square wave, one would expect to see a 1 kHz fundamental followed by the presence of a 3 kHz harmonic waveform, a 5 kHz harmonic waveform, 7 kHz harmonic waveform, etc. To carry out a Fourier Analysis, click the button highlighted in **Figure 6.31** in the active window with the plotted square wave.

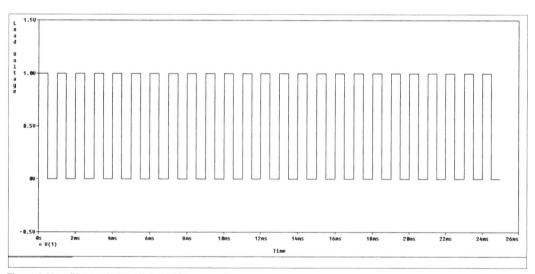

Figure 6.30 — Plot copied to clipboard in grayscale.

Figure 6.31 — FOURIER ANALYSIS button.

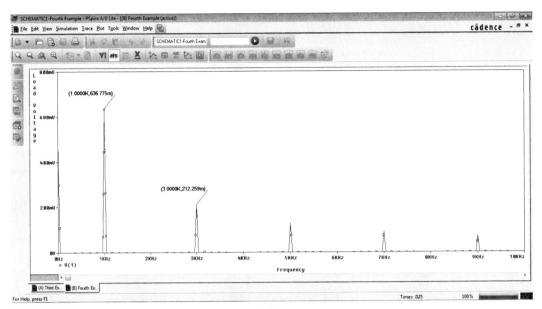

Figure 6.32 — Fourier Analysis, harmonic content of VPULSE square wave.

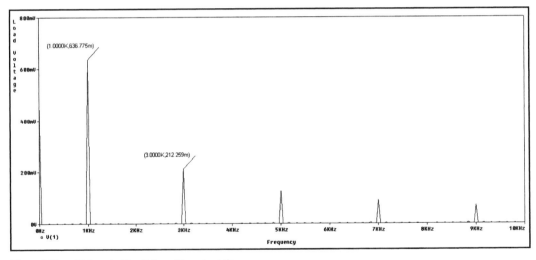

Figure 6.33 — Plot copied to clipboard in grayscale.

After clicking the FOURIER ANALYSIS button, you will be presented with the plot shown in **Figure 6.32**. The fundamental frequency and all of the associated harmonic frequencies, as well as their associated amplitudes, are plotted within the user-defined range of the X axis (Frequency Range).

The output is displayed just as was predicted. Finally, the output plot has been generated in a more legible grayscale plot in **Figure 6.33**.

In Summation

With the information from this chapter and the overview of the four basic simulation types from Chapter 4, you will be able to create basic designs using the Cadence software and present the data in the *Allegro AMS Simulator* window. In the next chapter, you will be introduced to semiconductor models that can be incorporated into more complex designs than those discussed thus far.

Chapter 7

Semiconductors

In previous chapters of this book, you were introduced to the basic schematic design process, passive components, and the fundamental simulation types that can be utilized within the *OrCAD/PSpice* environment. In this section we will expand upon all of these concepts while introducing you to another family of components featured in the *PSpice* library — semiconductors. In addition to the resistors, capacitors and inductors that you have already used to draw and run simulations, the *PSpice* library features a wide variety of diodes and transistors.

The Diode Model

Let us begin by drawing the circuit shown in **Figure 7.1**. The circuit consists of a single resistor, a dc voltage source and a 1N4002 diode. The 1N4002 diode is a relatively common rectifier diode. The default *PSpice* component libraries contain many different diodes ranging from a common variety to more obscure component models. An excellent feature of the *PSpice* software is that virtually every real component from every manufacturer has a *PSpice* model. They may not initially appear in the default *PSpice* component libraries, but they are available for download from most component manufacturers' websites or the Cadence/*PSpice* website. Each different diode model has its own characteristics and specifications that are indicated on the real components spec sheet.

To explore the behavior of these nonlinear components, the *PSpice* software can be used to plot their characteristic curves. To do so, one must perform a DC Sweep Analysis of the circuit. Varying the source voltage and plotting the current

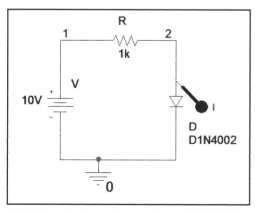

Figure 7.1 — Diode circuit using the model D1N4002.

in the circuit will provide a characteristic curve for this diode.

First, we need to declare the simulation parameters. **Figure 7.2** shows the SIMULATION SETTINGS window and defines the range of the DC Sweep starting at a value of –103 V and ending at +2 V with data points being plotted at 0.25 V intervals. After running the DC Sweep, you will generate a plot similar to the one shown in **Figure 7.3**.

Using this declared range for the DC Sweep provides a visual representation of the three modes of operation of this particular diode. Starting at the right of the plot, it is apparent that the diode is forward biased with a forward voltage of approximately 0.4 V. Continuing along the X axis, the trace of the characteristic curve extends past 0 V in the negative direction. This indicates that the diode is operating in reverse bias. As we continue farther in the negative direction of the X axis, the dc voltage source value eventually sweeps beyond the peak inverse voltage (PIV), and an increase of current flow in the negative direction can be noted as the diode is now operating in a state of breakdown. This seems to occur when the value of the voltage source V exceeds –100 V.

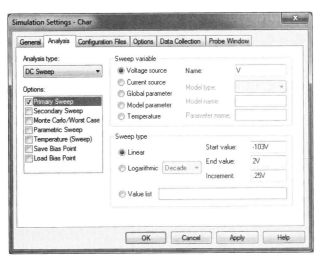

Figure 7.2 — Simulation settings for single DC Sweep.

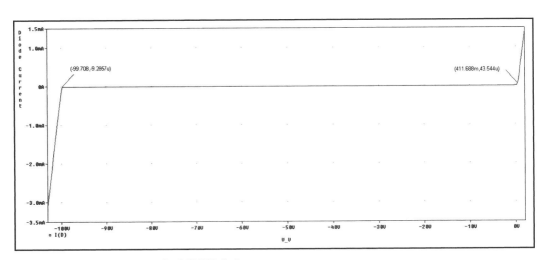

Figure 7.3 — Characteristic curve for D1N4002 diode.

Another simulation technique allows the user to provide a more accurate characteristic curve of a particular diode model by changing the axis variable of the X axis to the diode voltage. This will generate an output plot that more closely represents the characteristic curve of a diode as it may be seen on a spec sheet.

To accomplish this, open the AXIS SETTINGS menu from the Cadence *Allegro AMS Simulator* or *PSpice A/D* window that has the 1N4002 characteristic curve and select the X AXIS tab (see **Figure 7.4**). In this window, click the AXIS VARIABLE button to open the X AXIS VARIABLE menu. The X axis variable has been chosen as V(D:1), the voltage at the anode of the diode (see **Figure 7.5**).

Apply these changes and it will produce a plot representative of diode voltage versus diode current, as opposed to source voltage versus diode current. This is shown in **Figure 7.6.**

To further illustrate the difference between the two characteristic curves, the range of the DC Sweep has been extended in the positive direction to 10 V, and the X axis range of the diode voltage versus diode current plot has been defined from a minimum of –0.5 V to a maximum of 1 V. This will present the forward bias characteristics of the 1N4002 diode (see **Figure 7.7**).

Figure 7.4 — AXIS SETTINGS menu.

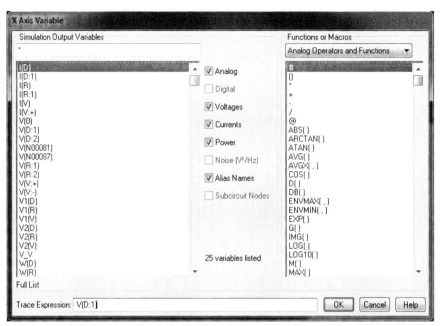

Figure 7.5 — X AXIS VARIABLE window.

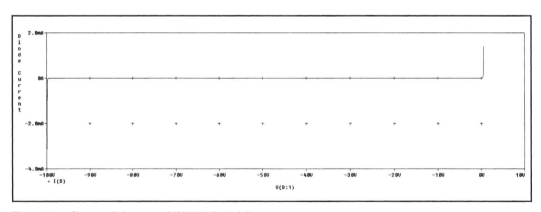

Figure 7.6 — Characteristic curve of 1N4002 diode, full range.

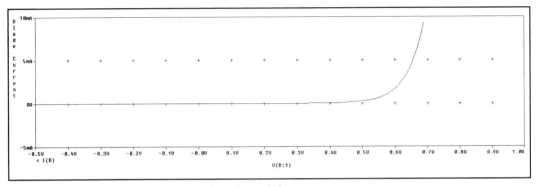

Figure 7.7 — Characteristic curve of 1N4002 diode, forward bias.

Bipolar Junction Transistors

We can build upon the concept of the basic DC Sweep simulation type when using the *PSpice* software to plot the characteristic curves for other types of semiconductors, specifically the transistor. Next we will look at both BJTs: the NPN and the PNP. To plot the characteristic curves of these three terminal semiconductors, we will be implementing two separate dc sources and performing two simultaneous DC Sweeps.

The first of the two BJTs to be studied will be the NPN. For this particular simulation, we will be using the 2N3904 model from the *PSpice* library. The construction of the circuit to be simulated is shown in **Figure 7.8**. A dc current source is connected to the base of the transistor. Sweeping this current source varies the base current and will have a direct impact on the value of

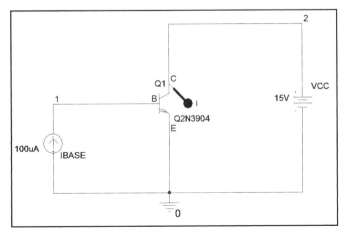

Figure 7.8 — NPN BJT test circuit schematic using the model Q2N3904.

the collector current due to the gain constant (β) of the transistor. Sweeping the voltage source will vary the voltage drop across the transistor. To plot the characteristic curves of this transistor, a current marker is placed on node C, the collector of the transistor.

Again, we must define the parameters of our simulation profile. For this type of simulation it is necessary to sweep two parameters of the circuit instead of one, as we are accustomed to from prior examples. Select the DC SWEEP analysis type, and inside the option section of the window ensure that the boxes are checked for both PRIMARY SWEEP and SECONDARY SWEEP. By default, PRIMARY SWEEP should already be checked. Using the mouse cursor, click/highlight PRIMARY SWEEP to display the primary sweep parameters. The primary sweep parameters will define the X axis of the output plot. In the case of the transistor characteristic curve, it is desirable to plot transistor current with respect to the change in V_{CE} (voltage across the transistor). Therefore, our dc voltage source, VCC, will be the source for the primary sweep. These parameters have been defined in **Figure 7.9**. The primary sweep parameters will remain the same for the rest of the examples in this chapter. The dc supply voltage connected across each transistor circuit will sweep from 0 V to 15 V, plotting data in 0.015 V increments.

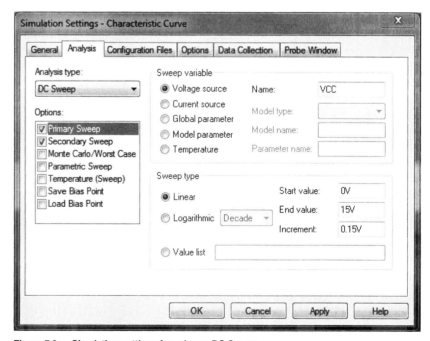

Figure 7.9 — Simulation settings for primary DC Sweep.

Now select SECONDARY SWEEP to declare the simulation parameters for our dc current source. For this simulation, the dc current source IBASE will vary from 0 A to 100 µA in incremental steps of 20 µA. Therefore, there will be separate traces on the generated output plot for each incremental step of IBASE (see **Figure 7.10**).

After pressing OK and running the simulation, a plot will be displayed similar to the one shown in **Figure 7.11**. The Y axis has been titled "Collector

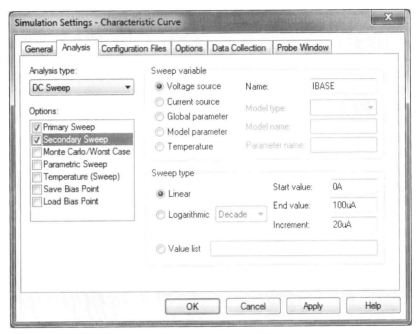

Figure 7.10 — Simulation settings for secondary DC Sweep.

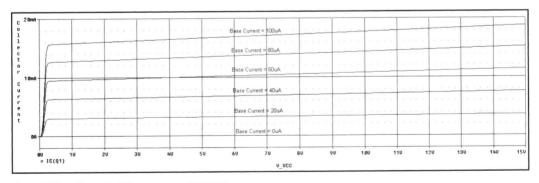

Figure 7.11 — Characteristic curves for Q2N3904.

Current" and each current trace has been labeled with its associated base current. This is a typical output plot for an NPN BJT. The 0 µA trace at the bottom of the plot indicates that the transistor is in cutoff. With IBASE = 0 µA, there is no base current to forward bias the base-to-emitter junction, therefore there is no collector current since the collector-emitter junction has an enormous internal resistance at this time. As the base current increases, an increase in collector current can be observed. This relationship will be maintained until the maximum base current and supply voltage have been reached, at which time the transistor is in saturation.

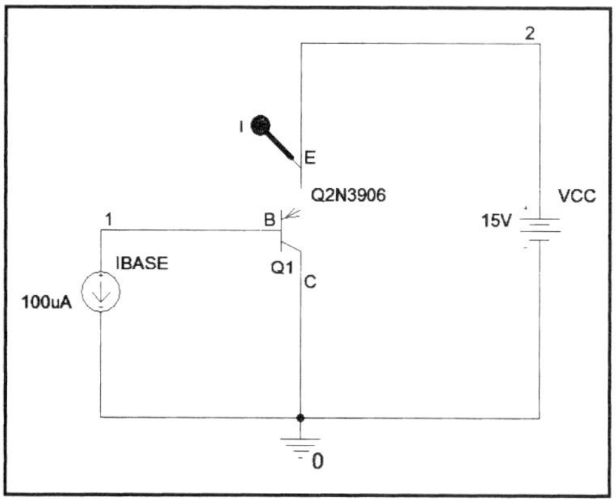

Next, the same simulation will be performed on the PNP BJT. This transistor relies on the same principles of operation as the NPN, the only difference being the physical makeup. The configuration of the N-type and P-type materials are inverted in the two transistor models, hence the names NPN and PNP. To compensate for this difference in their composition, our schematic design will vary slightly from the last exercise. Note that in comparison to the previous design, the transistor used in the design of the circuit in **Figure 7.12**, a 2N3906 (PNP) has been inverted

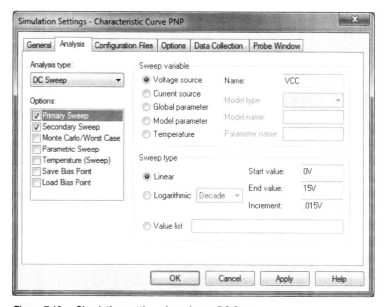

Figure 7.12 — PNP BJT test circuit schematic using the model Q2N3906.

Figure 7.13 — Simulation settings for primary DC Sweep.

with respect to VCC to ensure that the emitter now has a greater positive potential than the collector. Also, our current source, IBASE, has been reversed in order to forward bias the base-to-emitter junction. The BJT in this design shows a small gap in the path at the emitter terminal. There is a complete path at each at this node, but it doesn't happen to be shown with this particular *OrCAD* schematic symbol. Rarely, you will notice discrepancies like this with certain components in the *PSpice* library.

As can be seen in both **Figure 7.13** and **Figure 7.14**, the same simulation parameters have been used in this example that were used with the NPN model. Note that you could have kept the same schematic layout from the previous example and used it in this example with the 2N3906 and achieved the desired result. You would, however, need to modify the simulation parameters by defining them with negative magnitudes in the SIMULATION SETTINGS window, so instead of sweeping from 0 V to 15 V you would be sweeping from 0 V to –15 V.

This simulation produces almost an identical output plot (see **Figure 7.15**) to the one shown in Figure 7.11. The plot of the emitter current curves seen in Figure 7.15 are typical of a BJT. As the magnitude of the base current increases, the collector/emitter current will increase.

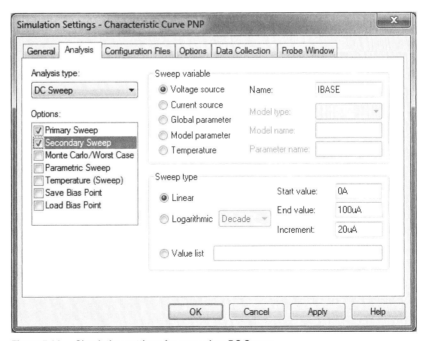

Figure 7.14 — Simulation settings for secondary DC Sweep.

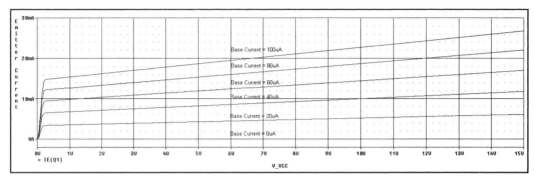

Figure 7.15 — Characteristic curves for Q2N3906.

Junction Field Effect Transistors

This same multiple-sweep procedure can be used to plot the characteristic curves for other types of transistors. Next we will look at another type of transistor — the junction gate field effect transistor (JFET). Curves will be simulated for both the N-Channel JFET and P-Channel JFET.

Figure 7.16 features a 2N5486 N-Channel JFET and two dc voltage sources. As in the previous example, this JFET shows small gaps in the path at both the drain and source terminals. There is a complete path at each of these nodes.

One dc source is the supply voltage to the drain, VDD, and the other will be swept to vary the gate voltage, VGG. As the gate voltage of a JFET varies, a change in current can be observed across the channel of the transistor

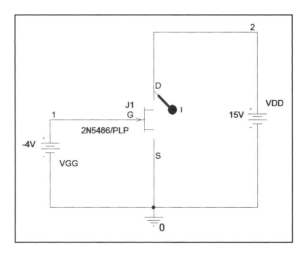

Figure 7.16 — N-Channel JFET test circuit schematic using the model 2N5486/PLP.

that exists between the source and drain terminals. The gate voltage will be adjusted in incremental steps similar to the way that the base current was modified in the case of the BJT. The voltage across the transistor will sweep at the same time to produce the set of characteristic curves for the JFET.

As previously stated, the supply voltage will be our primary sweep, and will range from 0 V to 15 V, with data points plotted in 0.015 V incremental steps (see **Figure 7.17**).

The secondary sweep of this simulation profile will be that of the gate voltage (see **Figure 7.18**). The voltage at the gate of the transistor will vary from –4 V to 0 V in 1 V increments. This is an N-Channel JFET, meaning that the voltage source VGG is connected to a P-type material. Applying a negative potential to the gate will widen the depletion region between this P-type gate, and the N-type channel, thus restricting current flow across the channel of the transistor.

After running the simulation, a plot will be generated similar to the one shown in **Figure 7.19**. Each of the drain current traces have been labeled with their associated value of gate voltage. As projected, when the gate voltage is 0 V, current through the channel is unrestricted. As the gate voltage increases in the negative direction, the channel current decreases.

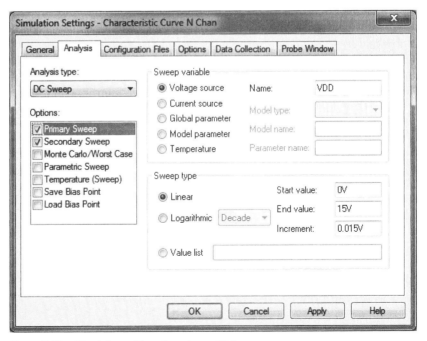

Figure 7.17 — Simulation settings for primary DC Sweep.

The current decreases as a direct result of the enlarging depletion region until it completely closes off the channel. The voltage across the gate that will yield no current flow across the channel is referred to as the pinch-off voltage. Looking at the plot below, the trace for VGG = –4 V either is or exceeds the pinch-off voltage. It is apparent with this trace that the channel has been closed completely since the drain current is constant at 0 amperes.

Next we will plot the characteristic curve for a J175 P-Channel JFET. **Figure 7.20** is very similar to the layout of Figure 7.16 but varies slightly in

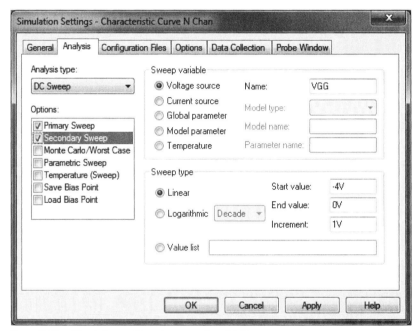

Figure 7.18 — Simulation settings for secondary DC Sweep.

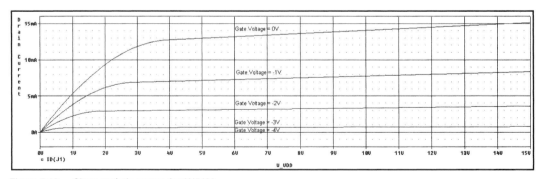

Figure 7.19 — Characteristic curves for 2N5486.

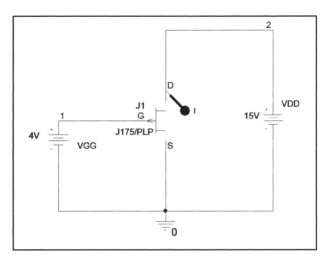

Figure 7.20 — P-Channel JFET test circuit schematic using the model J175/PLP.

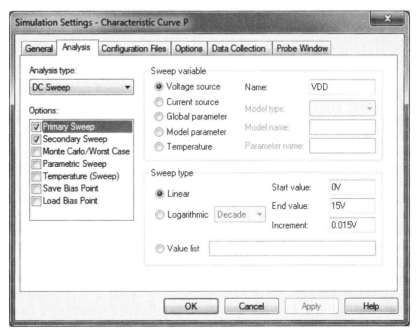

Figure 7.21 — Simulation settings for primary DC Sweep.

the same manner that the circuit was modified from the example of the NPN BJT to the PNP BJT. The modifications to the circuit shown in Figure 7.20 are to compensate for the physical construction of this particular transistor model. This is a P-Channel JFET. That means that the voltage source VGG is connected to an N-type material. Applying a positive potential to the gate will widen the depletion region between this N-type gate and the P-type channel, thus restricting current flow across the channel of the transistor.

The orientation of voltage source VDD has been changed from the N-Channel design to accommodate the P-Channel transistor model, so again, we will be sweeping from 0 V to +15 V, with data points marked in 0.015 V intervals (see **Figure 7.21**).

Figure 7.22 illustrates the simulation parameters of VGG, the voltage source applied to the gate. VGG will sweep from 0 V to 4 V in 1 V incremental steps. This will provide an output plot consisting of five channel currents, one for each gate voltage.

With a current marker placed at the source terminal it is evident that this P-Channel JFET exhibits similar behavior to the N-Channel JFET. According to the output plot generated in **Figure 7.23**, with a potential of 0 V applied to the gate the channel current is unrestricted. As the potential at the gate increases, the channel is eventually pinched off.

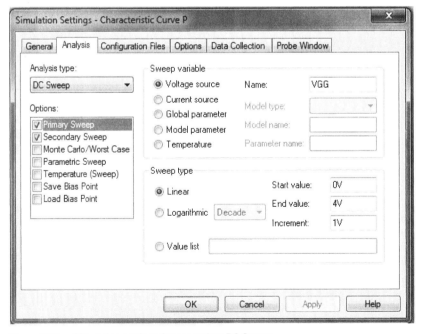

Figure 7.22 — Simulation settings for secondary DC Sweep.

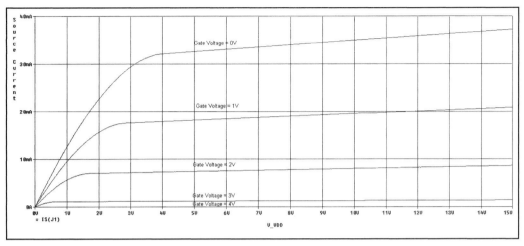

Figure 7.23 — Characteristic curves for J175.

Metal-Oxide-Semiconductor Field-Effect Transistors

The last type of transistor that will be covered is the metal-oxide-semiconductor field-effect transistor, or MOSFET. For this last example, we will use a relatively common N-Channel 2N7000 low-power switching MOSFET. This is an enhancement MOSFET, which will be tested using the schematic design shown in **Figure 7.24**. These enhancement MOSFETs differ in construction from a traditional JFET due to the configuration of the gate and the presence of an inversion layer comprised of SiO_2. With the N-Channel MOSFET, both the drain and the source connect to two isolated N-type materials that are separated by a larger quantity of a P-type

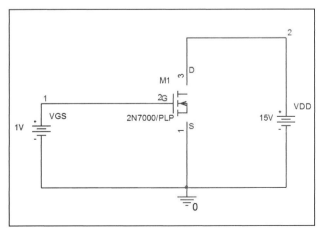

Figure 7.24 — N-Channel enhancement MOSFET test circuit schematic using the model 2N7000.

material. The gate connects to the SiO_2 layer instead of either of the doped semiconductor materials. With an N-Channel MOSFET, as positive potential builds at the gate, it repels the net positive charge of the P-type material, creating a channel between the two previously isolated N-type materials and allowing for current flow between the drain and the source.

As with all the previous examples, the supply voltage VDD will be the primary sweep and will range from 0 V to 15 V, marking data points in 0.015 V intervals (see **Figure 7.25**).

The parameters of the secondary sweep define the behavior and the range of voltage source VGS, the voltage applied to the gate. This voltage source will start at 0 V and adjust toward +5 V in 1 V increments. These parameters are shown in **Figure 7.26**.

The characteristic curves for this MOSFET are shown in **Figure 7.27**. Referring to the bottom traces in the plot for VGS = 2 V, 1 V, & 0 V, the potential at the gate was not sufficient enough to create a depletion region in the P-type material and allow current to flow between the N-type terminals. When VGS was increased to 3 V, the potential was significant enough to create a channel for current flow, and that channel only grew as VGS increased, as evident in traces VGS = 4V and VGS = 5V.

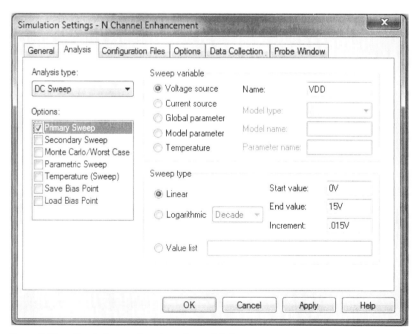

Figure 7.25 — Simulation settings for primary DC Sweep.

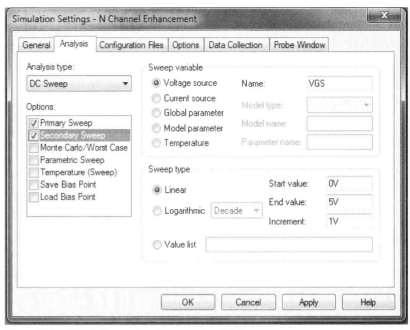

Figure 7.26 — Simulation settings for secondary DC Sweep.

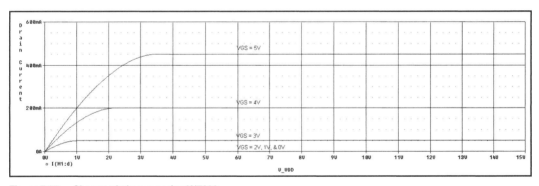

Figure 7.27 — Characteristic curves for 2N7000.

Design Problem

For the last example in this chapter, the transistor can be demonstrated in a practical circuit while we build further upon the concept of the DC Sweep simulation profile in a design application. Assume that we are trying to design an audio amplifier represented in the schematic design of **Figure 7.28**. This circuit utilizes the 2N3904 NPN BJT seen in the second example in this chapter in a common emitter configuration. The potential established between the voltage divider of R1 and R2 will be used to bias this basic transistor amplifier circuit. Without the capabilities of this kind of software, you would have to sit down with spec sheets and scratch paper and carry out numerous calculations to determine how to properly bias this amplifier or one like it.

The Cadence software allows a user to sweep not only the values of a voltage or current source, but individual component values as well. This can be especially helpful in the instance of trying to bias an amplifier such as the one in Figure 7.28.

One can begin by constructing the schematic design shown using the *OrCAD* software. Instead of entering a value for R2, the bottom resistor of our voltage divider, enter "{RVAL}". Next, search the Component library for the component PARAM and place it anywhere on the schematic page.

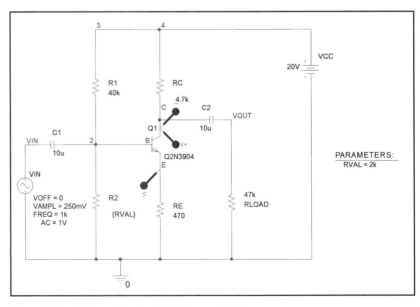

Figure 7.28 — BJT amplifier circuit with sweepable R2.

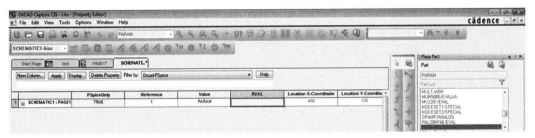

Figure 7.29 — PROPERTY EDITOR for PARAMETERS component, RVAL column has been selected to be modified.

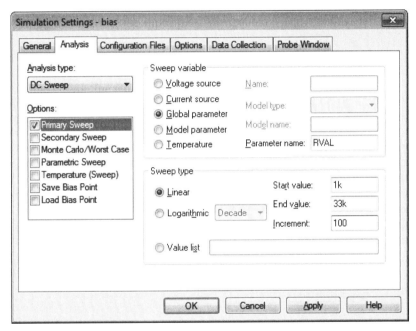

Figure 7.30 — Sweep characteristics for the RVAL parameter.

Double-click on the newly placed parameter component to display its properties.

Next search the Component library for the component PARAM and place it anywhere on the schematic page. Double-click on the newly placed PARAMETERS: component to display its properties, as seen in **Figure 7.29**. Create a new column titled "RVAL", the value you declared for R2. For this property, you can insert any value of resistance that you desire, and if you were to run a Transient Analysis, this is the value that you would use. For

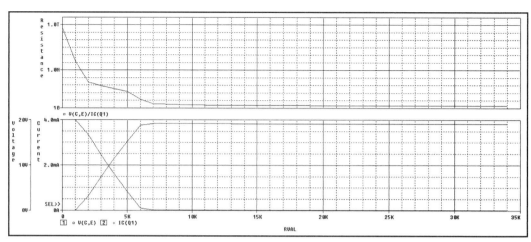

Figure 7.31 — *PSpice* output displaying effects of the sweep on the BJT amplifier.

the purpose of the DC Sweep this value will not be considered. For now a value of 2 kΩ has been entered.

For the purpose of this example, a value of R2 needs to be established to properly bias the amplifier. This can be determined by measuring the collector current and collector-emitter voltage while sweeping the value of R2. To accomplish this, create a DC Sweep simulation profile (see **Figure 7.30**). In the SWEEP VARIABLE portion of this window, select GLOBAL PARAMETER, and for a parameter name enter "RVAL", the same value that you declared for the component R2. This GLOBAL PARAMETER sweep function refers to the PARAMETERS: component placed on the schematic page. Remember that the PARAM component is associated with component R2 using the declared value "RVAL" and will sweep the value of this resistor based on the limits declared in the bottom portion of this window. According to these constraints, R2 will sweep from 1 kΩ to 33 kΩ in 100 Ω increments.

After running the simulation, plots resembling **Figure 7.31** are generated as a result of the voltage and current markers in the schematic design. They have, of course, been manipulated slightly in the *PSpice* PROBE window to make them more legible. A BJT can be thought of conceptually as a voltage-controlled resistor. The top plot illustrates this point as a trace expression and has been entered as "V(C,E)/IC(Q1)" to plot the internal resistance of the transistor. As the value of R2 increases, the base voltage increases and the resistance of the transistor drops. As this change occurs, observe the bottom plot. The bottom plot consists of traces plotting both collector current and collector emitter voltage with respect to the value of R2. As the resistance of

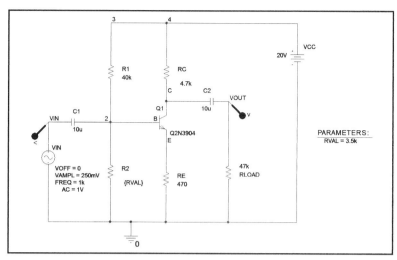

Figure 7.32 — BJT amplifier schematic with new R2 value.

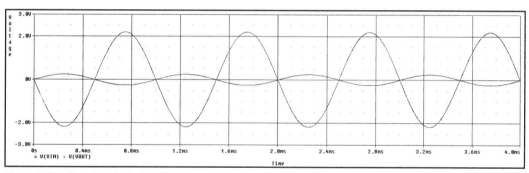

Figure 7.33 — Transient Analysis with R2 set at 3.5 kΩ.

the transistor drops, the collector current increases.

At the beginning of the sweep, and soon after R2 begins to change, the transistor is in cutoff. Looking at the plot it is easy to find a value of resistance that is right in the middle, in a linear mode of operation, between a state of cutoff and saturation. That value is a 3.5 kΩ.

With this newly obtained value of resistance, you can refer back to the original design (see **Figure 7.32**). Inside the parameter component on the schematic page, change the value of resistance to 3.5 kΩ and create a new simulation profile for a Transient Analysis. Place voltage markers at the input and output of the circuit and run the simulation.

Figure 7.33 plots the output voltage alongside the input voltage and demonstrates the proper bias of this transistor circuit using a value of 3.5 kΩ for R2. This choice for a resistor value yields a clean output with the proper headroom. This was done without any additional freehand computation and further showcases the comprehensiveness of the *OrCAD/PSpice* software.

In Summation

Now that you have an introduction to semiconductor components, you can start to take better advantage of what the *OrCAD/PSpice* software has to offer by designing and simulating circuits with practical applications. These types of circuits are discussed further in Chapter 11.

Chapter 8

Miscellaneous Components

Up to this point, we've touched upon the most basic passive components and some of the semiconductor models available in the *PSpice* software. Before moving into sample circuits and other advanced operations of the software, let's discuss some of the other component models that are available. The components to be discussed in this chapter will be switches, op-amps, and transformers. We'll begin with switches.

Switches

There are two types of switches are available in the *PSpice* library — Sw_tClose and Sw_tOpen. The letters "Sw" denote that the component being used is a switch. These switch components can be used only with the Transient Analysis simulation type. Their operation depends on a value of time that is declared by the user for when they will either open or close. The component Sw_tClose is a switch that is initially open, but will close at a time that is declared by the user. This type of switch is utilized in the schematic design shown in **Figure 8.1**.

Figure 8.1 is a simple series circuit with a 10 V source connected to a 1 kΩ load resistor. A Sw_tClose switch component has been placed in the series path between nodes 1 and 2. As stated, this switch is open and will close depending on the user-defined parameter of time. Referring again to the schematic of Figure 8.1, this parameter is titled TCLOSE and

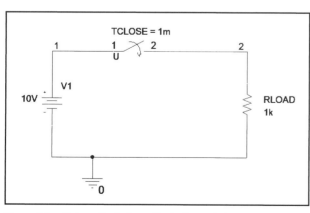

Figure 8.1 — Schematic design with closing switch.

has been set at 1 ms. A Transient Analysis simulation profile has also been set up to run for a total of 2 ms starting at 0 s with a step size of 2 μs (see **Figure 8.2**).

The load voltage has been measured during this Transient Analysis simulation and plotted in the probe graph shown in **Figure 8.3**. This is a result of placing a voltage marker on node 2 of the schematic design. From 0 to 1 ms there is no load voltage present. The switch is initially open, so there is no current flow through the series path. At 1 ms, the switch closes,

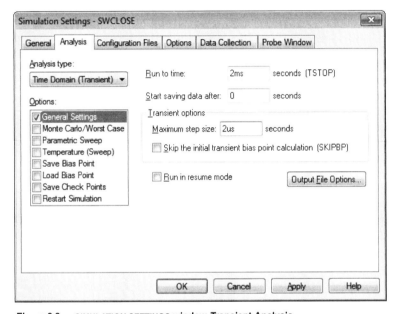

Figure 8.2 — SIMULATION SETTINGS window, Transient Analysis.

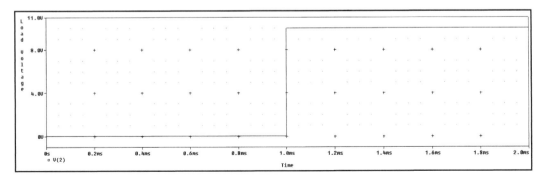

Figure 8.3 — *AMS Simulator* window measuring load voltage as the switch closes.

and from 1 ms to 2 ms the 10 V dc source voltage drops across the load resistor.

Conversely, the component Sw_tOpen is a switch that is initially closed. It will open at a time that is declared by the user. This timing parameter is called TOPEN. To illustrate how these switches can modify the behavior of the most basic circuits, both Sw_tOpen and Sw_tClose will be used in the

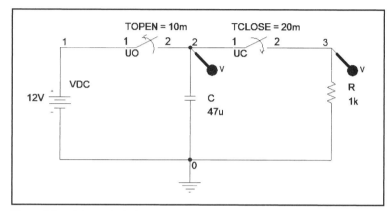

Figure 8.4 — Schematic design using both opening and closing switches.

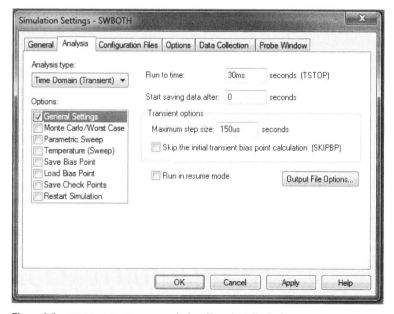

Figure 8.5 — SIMULATION SETTINGS window, Transient Analysis.

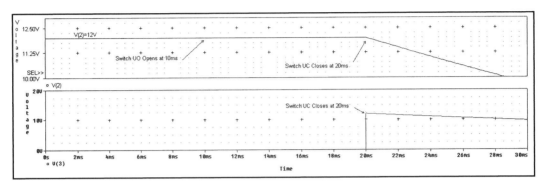

Figure 8.6 — *AMS Simulator* window measuring node voltages as the switches open and close.

simple parallel RC circuit shown in **Figure 8.4**. The switch that is initially closed lies between nodes 1 and 2 and will open at 10 ms, while the switch that is initially open lies between nodes 2 and 3 and will close at 20 ms.

A Transient Analysis will be carried out on this schematic input file. It will start at 0 s and run for a duration of 30 ms with a defined step size of 150 μs. These simulation parameters have been defined in the SIMULATION SETTINGS window shown in **Figure 8.5**.

The top plot of **Figure 8.6** is a measure of the voltage from node 2 with respect to ground. The bottom plot depicts the change in voltage measured from node 3 with respect to ground as this Transient Analysis takes place.

Let's refer to the schematic design shown in Figure 8.4. The switch UC, Sw_tClose is initially open for the first 20 ms of the simulation, and the switch UO, Sw_tOpen is closed for the first 10 ms of the simulation. For the first 10 ms of the simulation, the 12 V dc source is connected only across the 47 μF capacitor. The capacitor charges almost instantaneously, and after 10 ms, the switch UO opens. At this time, the source VDC is no longer connected and the switch UC has not yet closed, so the capacitor remains charged until 20 ms. At 20 ms, the switch UC closes, and the capacitor begins a steady discharge across the 1 kΩ resistor that will last a period of 5 τ (235 ms) that exceeds the range of the Transient Analysis.

Operational Amplifiers

As you can see, switches can provide great versatility to the operation and functionality of a circuit. They can aid its operation by either turning a circuit on and off, or they can add additional branches that will modify a circuit's behavior during a Transient Analysis.

Figure 8.7 features a model of the common uA741 operational amplifier. Even in the *OrCAD/PSpice* software environment, the schematic

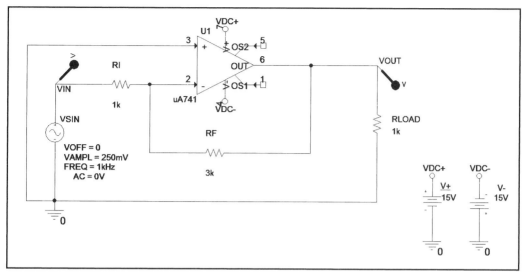

Figure 8.7 — Op-amp schematic using an inverting configuration.

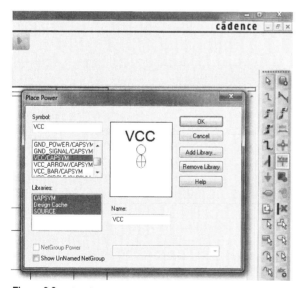

Figure 8.8 — PLACE POWER button highlighted and PLACE POWER window active displaying VCC terminal component.

symbol maintains the typical pinout for a uA741 package. Pins 1 and 5 are the offset null pins, pins 2 and 3 are the inverting and non-inverting inputs, respectively, pins 4 and 7 are used for power, and pin 6 is the output. For the purpose of this example, the usage of the uA741 will be demonstrated with an inverting configuration. The circuit has an input of 250 mV at 1 kHz.

This schematic design uses power terminals to connect dc voltages to pins 4 and 7. These terminals come in especially handy when working with complex designs or circuits that use multiple voltages. Placing a voltage source schematic symbol numerous times in a schematic design or placing a voltage source schematic in a confined space can make your design appear messy. In the toolbar to the right side of the *OrCAD* interface, click on the highlighted icon to display the PLACE POWER window. This is shown alongside an active PLACE POWER window in **Figure 8.8**.

The VCC component can be used as a reference terminal for power in the schematic design. In terms of the schematic design shown in Figure 8.7, it has been used a total of four times. Because of the feedback loop from pin 6 to pin 2, it would have been too cluttered to place a VDC component and ground symbol in that small space. A VCC terminal was placed on pin 4 and was renamed VDC–. A second VCC terminal was placed in an open space in the schematic design and given the same name of VDC–. This floating (not connected) VDC– terminal was then connected to the negative side of a 15 V dc source. The positive side of the voltage source was grounded. This creates an indirect schematic path between pin 4, the negative voltage source, and ground. The same process was repeated with the second pair of VCC terminals using a reference name VDC+, indirectly attaching a positive 15 V dc source to pin 7 of the operational amplifier. Assuming that we are working with a larger circuit consisting of multiple op-amps, these terminals VDC+ and VDC– can be used repeatedly, anytime these positive or negative 15 V potentials are required.

A Transient Analysis is performed on the circuit of Figure 8.7. These simulation parameters are shown in **Figure 8.9**. The simulation will start at 0 s and end at 2 ms with a step size of 2 μs.

The feedback resistor, RF, in the inverting op-amp circuit is 3 kΩ and

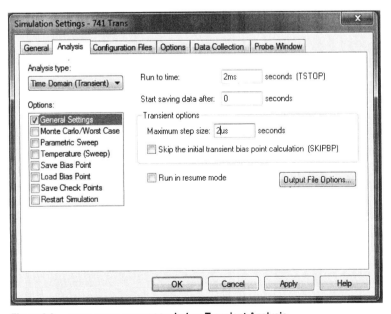

Figure 8.9 — SIMULATION SETTINGS window, Transient Analysis.

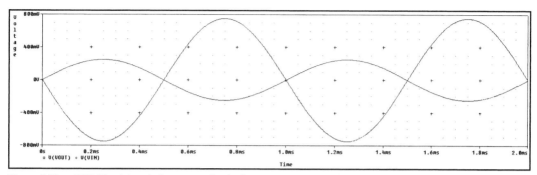

Figure 8.10 — *AMS Simulator* window measuring both output and input voltage of op-amp circuit.

the input resistor, RI, is 1 kΩ. Therefore, a calculated gain is expected to be –3. With the 250 mV input ac voltage, this will yield an output waveform with an amplitude of 750 mV that is 180° out of phase with the input signal. Looking back at the input schematic file (Figure 8.7), voltage markers have been placed near the ac source at net alias VIN and across the load resistor at the net alias VOUT. The plot in **Figure 8.10** shows a measurement of voltage at net aliases VIN and VOUT. The relationship between the input and output voltage is exactly as expected in terms of phase relationship and amplitude.

Transformers

The last topic of discussion in this chapter is the transformer. *PSpice* features many different transformer models, both linear and nonlinear, with multiple primaries, multiple secondaries, center taps, and with different core materials. For the purpose of this example, we will use a basic linear transformer with a single primary and single secondary (see **Figure 8.11**).

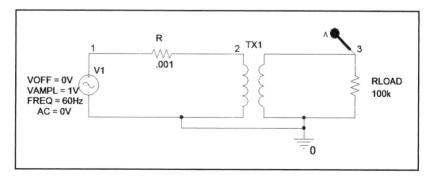

Figure 8.11 — Linear transformer circuit schematic.

The primary is connected to a 1 V ac source and the secondary is connected across a load. The frequency of the ac source is 60 Hz.

To explore the parameters of the transformer that can be modified, double-click on the component TX1 to open its PROPERTY EDITOR (see **Figure 8.12**). For the basic linear transformer, the main parameters are COUPLING, L1_VALUE, and L2_VALUE.

The numerical value in the COUPLING column of the PROPERTY EDITOR is representative of the coupling coefficient of a transformer. The coupling coefficient ranges from 0 to 1 and is essentially a ratio of how much of the magnetic flux from the primary reaches the secondary. In regard to a physical

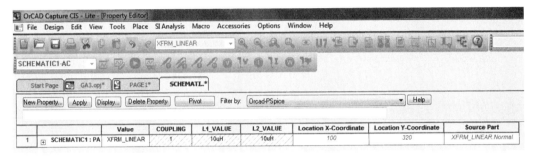

Figure 8.12 — PROPERTY EDITOR window **XFRM_LINEAR** component.

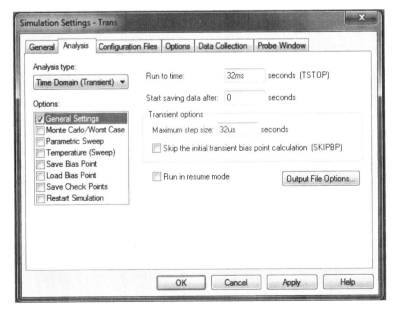

Figure 8.13 —
SIMULATION
SETTINGS window,
Transient Analysis.

transformer, a transformer is tightly coupled if the primary and secondary coils are in close proximity to each other and a majority of the flux from the primary reaches the secondary. Transformers have loose coupling if the coils are distant and little of the flux from the primary reaches through the secondary. Tightly coupled transformers have a coupling coefficient that is close to 1. In this example, we are dealing with an ideal transformer, so the coupling coefficient has been declared as 1.

L1_VALUE and L2_VALUE are the inductive properties of what are the primary and secondary windings in this circuit. For this example, the inductance of both the primary winding and secondary winding have been set at 10 µH.

A Transient Analysis will be carried out on this circuit that will start at 0 s and end at 32 ms with a step size of 32 µs as shown in the SIMULATION SETTINGS window in **Figure 8.13**. This run time has been selected to display an output plot that executes just under two complete cycles of the output waveform.

With the voltage marker placed at node 3, the output plot (see **Figure 8.14**) displays the output voltage that is measured across the 100 kΩ resistor connected to the secondary winding of the transformer. Since this transformer is an ideal linear transformer with a coupling coefficient of 1, the amplitude of the voltage across the secondary is equal to the primary/source voltage.

Lastly, we can use an AC Sweep simulation profile to measure the frequency response of the transformer. Since we are working with a linear transformer, it can be expected that a rather consistent output will be measured throughout the frequency spectrum. **Figure 8.15** displays the simulation parameters for the AC Sweep. The logarithmic sweep will begin at 10 Hz and end at 1 MHz with 100 data points per decade.

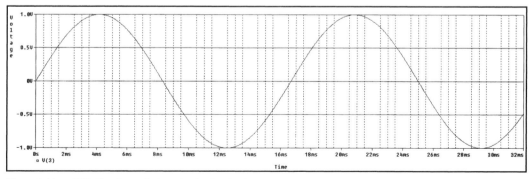

Figure 8.14 — *AMS Simulator* window plotting load voltage with respect to time.

Again, the voltage marker has been placed at node 3 to measure the secondary voltage in the probe graph shown in **Figure 8.16**. The output is characteristic of a linear transformer and features a flat frequency response throughout the range of the sweep.

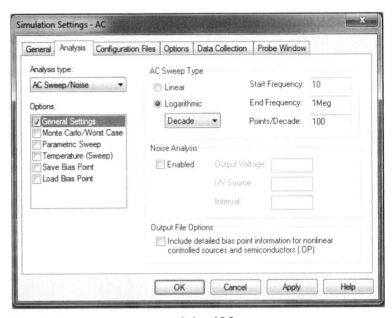

Figure 8.15 — SIMULATION SETTINGS window, AC Sweep.

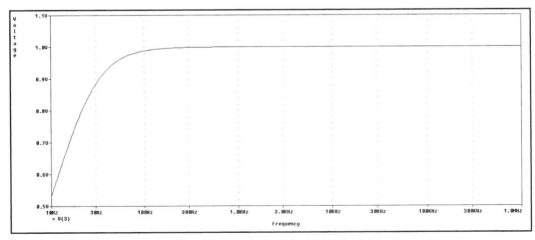

Figure 8.16 — *AMS Simulator* window plotting load voltage with respect to frequency.

In Summation

These components can only further enhance the capabilities of the basic designs discussed thus far. Chapter 11 will provide an in-depth examination of circuits with practical applications and will showcase many different types of sample circuits using them.

Chapter 9

Transmission Lines

If you are using *PSpice*, the basic transmission line element is actually considered to be lossless. This is quite reasonable for a circuit analysis program with integrated circuit emphasis, since transmission lines used within an integrated circuit (likely to be stripline or microstrip) are going to be very short, both physically and electrically. The loss in a short transmission line is normally quite small (a fraction of a dB) and, therefore, negligible for most purposes.

If you wish to add loss to a *PSpice* transmission line, you can do so by using lumped (as opposed to distributed) resistive elements. The dominant loss mechanism in most practical transmission lines is the conductor loss due to skin effect, not the dielectric loss. You can simulate a long and lossy line by breaking it up into shorter sections and adding a small resistance in series with each short section to represent the distributed conductor loss. Although it is tempting to use extremely short sections (each with a very small series R for loss), there is a tradeoff involved with this. In Transient Analysis, *PSpice* will make the timestep (or internal computing interval) less than or equal to one-half the minimum transmission delay of the shortest transmission line. This can make a *PSpice* Transient Analysis take a very long time if short transmission lines are used. Another problem is that transmission line loss increases with frequency. It is not possible to specify a frequency-dependent resistor, so line loss simulated by lumped resistors will be valid at only one frequency. The *PSpice* limitation that transmission lines are lossless is not a serious one. In the examples that follow, you will see that *PSpice* can be used to solve an assortment of transmission line problems, which would be all but impossible to solve in a reasonable time without a computer.

Transmission Line Length

Since there are two methods for telling *PSpice* the length of a transmission line, it may be helpful to look at how each method works and how to convert from one method to the other.

In general, we can say that the transmission delay, TD, of a transmission line can be found from:

TD = Physical Length/Velocity
where Velocity is the phase velocity in the line, given by:
Velocity = (Velocity Factor) × (Free-Space Velocity)
The Velocity Factor is less than or equal to 1.00.

Also, the normalized electrical length, NL, is:
NL = Physical Length/Wavelength
where Wavelength is the wavelength in the line, determined by:
Wavelength = Velocity/Frequency

Normalized electrical length can then be expressed as:

NL = (Physical Length × Frequency)/Velocity
Since Physical Length = TD × Velocity = (NL ×Velocity)/Frequency, then:
TD = NL/Frequency
and
NL= Frequency × TD

Let's apply this information to a practical example. Say we have a transmission line that has a characteristic impedance of 50 Ω, is 12 meters long, and has a velocity factor of 0.66. Now let's ask:
A) What is the transmission delay?
B) What is the normalized electrical length?
We know that:
A) Velocity = Velocity Factor × (c), and c = 3×10^8 m/s, so Velocity = $0.66 \times (3 \times 10^8$ m/s) = 2×10^8 m/s. TD = Length/Velocity = 12 m/(2×10^8 m/s) = 60 ns.

B) Before the normalized electrical length can be determined, a frequency must be specified. Let's use 30 MHz. NL = (Frequency × Length)/ Velocity = [(30×10^6) ×12]/2×10^8, so NL = 1.8 wavelengths.

Each of these characteristics of a transmission line can be modified using the PROPERTY EDITOR in the Cadence *Allegro Design Entry/ OrCAD Capture* schematic capture window. Each characteristic of a transmission line is abbreviated in the PROPERTY EDITOR as shown below:
- Design Frequency, F
- Normalized Electrical Length, NL
- Transmission Delay, TD

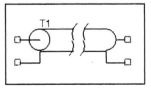

Figure 9.1 — Transmission line schematic symbol (T/ANALOG in the library).

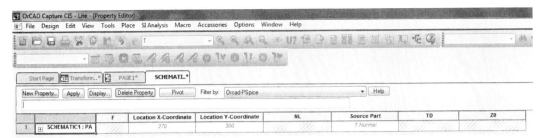

Figure 9.2 — PROPERTY EDITOR window for transmission line component.

These parameters will be adjusted in the example problems later in this chapter to manipulate the behavior of each transmission line circuit. **Figure 9.1** is the schematic symbol for transmission line in the schematic capture software (T/ANALOG in the library).

To modify each of these parameters, double-click on the transmission line schematic symbol to display the PROPERTY EDITOR. In the PROPERTY EDITOR, each of these characteristics can be modified to accommodate a particular design (see **Figure 9.2**).

As with all the components discussed in previous chapters, a transmission line component has a corresponding element line in a circuit's netlist. It is strongly recommended that users generate netlists of their own transmission line designs in an effort to develop a stronger understanding of the composition and purpose of a netlist.

Matched Load

This problem will show how *PSpice* can be used in the time domain to look at a single short pulse at the sending end and the receiving end of a physically short transmission line.

Figure 9.3 shows a pulse voltage source, with 50 Ω output impedance,

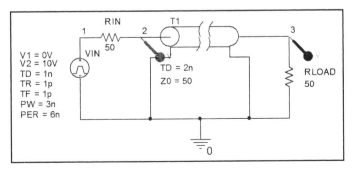

Figure 9.3 — Transmission line circuit schematic with matched load.

connected to a 50 Ω transmission line, which is terminated in a 50 Ω matched load. The characteristics of the pulse are defined in the schematic design. The initial voltage will be 0 V and will rise to 10 V after a 1 ns delay. The pulse will have a width of 3 ns and will have both a 1 ps rise time and fall time. More information can be gathered from the schematic design shown in Figure 9.3.

Transmission line T1 has a TD of 2 ns. We should, therefore, expect

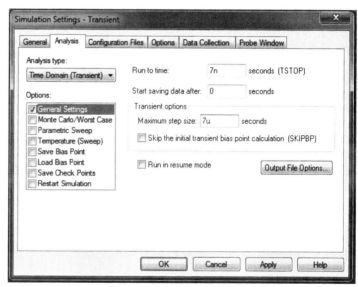

Figure 9.4 — SIMULATION SETTINGS window, Transient Analysis.

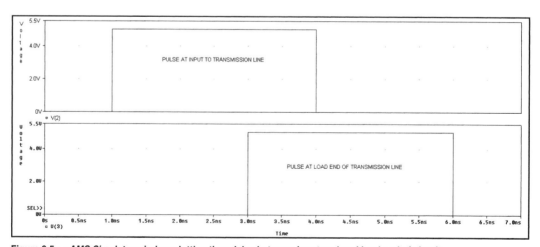

Figure 9.5 — *AMS Simulator* window plotting time delay between input end and load end of circuit.

that whatever voltage changes appear at the input to the transmission line (node 2) should occur 2 ns later at the output (node 3).

The *PSpice* SIMULATION SETTINGS window (see **Figure 9.4**) specifies that a Transient Analysis will run from 0 ns to 7 ns with a step size of 7 μs. The results of the *PSpice* Transient Analysis, displayed by PROBE, are shown in **Figure 9.5**. Voltage V(2) is shown in the upper plot, and the lower plot shows V(3).

The input voltage, V(2), rises from 0 V to 5 V at 1 ns and remains 5 V for 3 ns. The output voltage, V(3), does the same as V(2) except that it is delayed by 2 ns. Since the transmission line is terminated in a matched load, no reflections are seen. If the load is not matched, a reflection will occur at the mismatch, and the input voltage will change at a time equal to twice the transmission delay of the line. This is the basis of time-domain reflectometry, which is used at the input side to examine transmission lines for defects. *PSpice* can be used to predict what a time-domain reflectometry display should be for a certain transmission line circuit.

The last example in this chapter demonstrates time-domain reflectometry with mismatches between both generator/transmission line and transmission line/load.

Balun Connected to a Voltage Source

PSpice will be used to determine the power bandwidth of a transmission line balanced-unbalanced (balun) transformer. Baluns are used to connect coaxial transmission lines to balanced transmission lines or antennas, and utilize a ½ wavelength section of transmission line to accomplish the impedance transformation.

It is a very simple task to analyze how baluns work at the design frequency, where the line is indeed ½ wavelength long. However, it is not at all simple to analyze how the balun will perform at frequencies above or below the design frequency. *PSpice* can be used to great advantage to perform a frequency sweep of the input voltage and to plot the output voltage

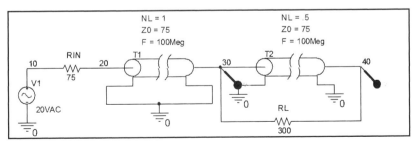

Figure 9.6 — Transmission line balun circuit schematic with voltage source.

versus frequency. We can tell the power bandwidth of the balun circuit by observing when the output voltage magnitude drops to 0.707 of its value (a drop of 3 dB) at the design frequency.

Look at the circuit shown in **Figure 9.6**. The 75 Ω coaxial transmission line impedance is converted by the balun to the 300 Ω impedance of the

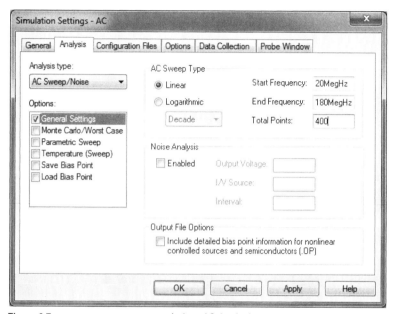

Figure 9.7 — SIMULATION SETTINGS window, AC Analysis.

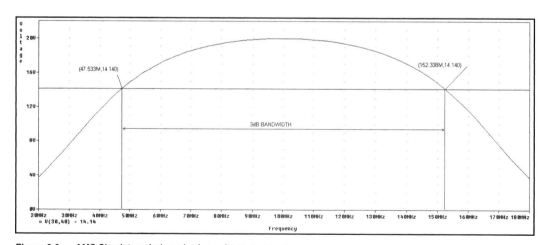

Figure 9.8 — *AMS Simulator* window plotting voltage across RL.

load. The design frequency of the balun is 100 MHz, so transmission line T2 is ½ wavelength long at 100 MHz. **Figure 9.7** shows the simulation profile to be used on this schematic design, an AC Analysis in which the frequency sweeps linearly from 20 MHz to 180 MHz.

The PROBE display of the *PSpice* analysis output data (**Figure 9.8**) shows the graph of the voltage magnitude across the load resistor, trace V(30,40), to be a maximum of 20 V at 100 MHz. Using the 0.707 criterion, the cursors in PROBE show that the voltage falls to 0.707 × 20 V, or 14.14 V, at approximately 47.6 MHz and 152.4 MHz.

Thus, *PSpice* has directly provided the information to determine the power bandwidth to be about 152.4 MHz − 47.6 MHz, or 104.8 MHz. The power bandwidth is not necessarily the same as the usable bandwidth, since the voltage standing wave ratio (VSWR) may be acceptable only over a much narrower bandwidth. The higher the VSWR, the more power is reflected by the load back to the source. The next problem will show how to make *PSpice* provide information that can be used to calculate VSWR at each frequency.

A very simple modification can be made to a circuit to make *PSpice* calculate the input impedance of a circuit. In this next example, we will slightly change the circuit of the previous problem to allow PROBE to graph the input impedance of the balun circuit at each frequency. From this input impedance data, the VSWR at each frequency can easily be calculated, and a VSWR-based bandwidth can be determined.

Balun Connected to a Voltage Source

If an ac current source of 1 A magnitude and 0° phase is connected to the input of a circuit, the phasor voltage at the input is the product of current and impedance, as we see below:

$$V = I \times Z = 1.0 \times Z = Z$$

Thus, by replacing the ac voltage source and 75 Ω resistor in the previous problem with a 1 A current source, *PSpice* will generate input impedance data directly. Refer to the modified schematic diagram in **Figure 9.9**, which shows current source I1. Since the phase is not specified in the I1 current source, the default value of 0° will apply.

Figure 9.10 shows the simulation parameters for this circuit. A linear AC Sweep will occur from 60 MHz to 140 MHz with a total of 400 data points.

The PROBE output graph shown in **Figure 9.11** has two plots that show the complex input impedance in polar form — V(20), the magnitude of the voltage at node 20, and P(V(20)), the phase of the voltage at node 20. These are the magnitude and phase, respectively, of VIN.

The input impedance data, when converted into VSWR data, indicates that the VSWR at 60 MHz and 140 MHz exceeds 3. The conversion method can be found in many texts on electronic communications with a chapter on transmission lines. While the data at 48 MHz and 152 MHz was done in this analysis, the VSWR at those frequencies is 5.66. A VSWR that high would be unacceptable for most applications.

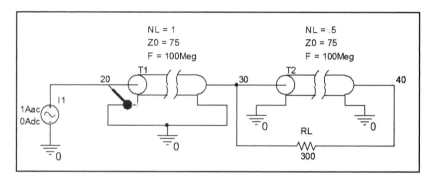

Figure 9.9 — Transmission line balun circuit schematic with current source.

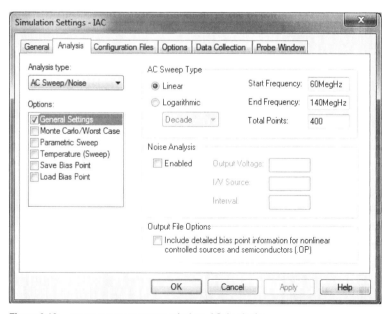

Figure 9.10 — SIMULATION SETTINGS window, AC Analysis.

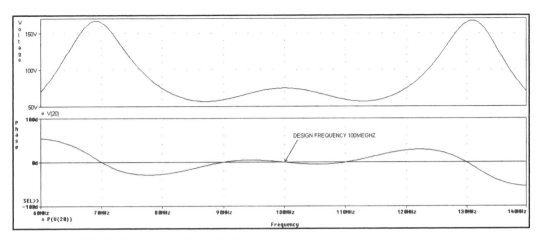

Figure 9.11 — *AMS Simulator* window plotting both output voltage and phase angle.

Quarter-Wavelength Transformer

One method of matching a load to a transmitter is to use a ¼ wavelength section of transmission line to transform the impedance at its load side to the characteristic impedance of the system. Refer to **Figure 9.12**.

The purpose of T3 (a 50 Ω line) is to transform the very low resistance of the load, 5 Ω, to 500 Ω at node 4. Transmission line T2 (not a 50 Ω line) is ¼ wavelength long and transforms the impedance of 500 Ω at node 4 to 50 Ω at node 3, thus creating a perfect match at the design frequency.

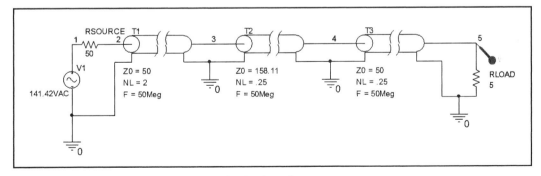

Figure 9.12 — Quarter wavelength transformer circuit schematic.

Figure 9.13 details the simulation settings for this circuit and calls for an AC Analysis. A linear AC Sweep will occur from 10 MHz to 90 MHz with a total of 400 data points. The results of the AC Analysis will be plotted in the *PSpice* PROBE window shown in **Figure 9.14**.

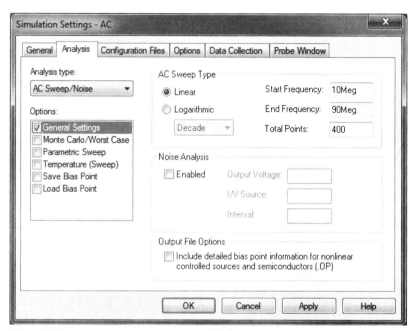

Figure 9.13 — SIMULATION SETTINGS window, AC Analysis.

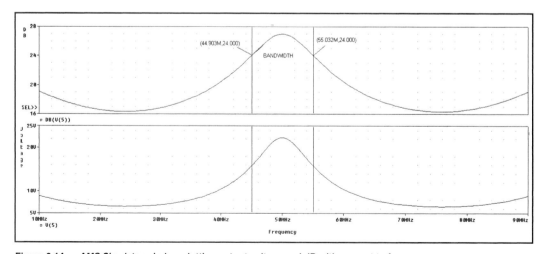

Figure 9.14 — *AMS Simulator* window plotting output voltage and dB with respect to frequency.

The voltage magnitude across the load, V(5), is plotted two ways in Figure 9.14. The lower plot is the magnitude of node 5 voltage, while the upper plot shows 20 log (node 5 voltage). The decibel data in the upper plot makes it easy to evaluate the half power, or –3 dB, bandwidth of this matching circuit. Using the cursors in the *PSpice* window, the –3 dB points have been indicated and designate the bandwidth.

The DB(V(5)) graph (upper plot) shows that load voltage is a maximum of 27 dBV at the design frequency of 50 MHz and falls off on either side. The 3 dB bandwidth is the range of frequencies over which the magnitude is within 3 dB of 27 dBV. The cursors in PROBE easily show that the bandwidth is from approximately 45 MHz to 55 MHz, a 10 MHz range.

Time-Domain Reflectometer

An excellent way to understand how pulses behave on transmission lines is to use a time-domain reflectometer (TDR), which sends a train of voltage steps into the sending end of a transmission line and displays the reflections when they arrive back at the sending end. By observing the sending-end voltage, with a knowledge of voltage reflection coefficients, you can quickly determine both the length of the transmission line and the nature of the load impedance. A similar process is used by bats (in air) and by porpoises (in water); it is called SONAR.

Ordinarily, the TDR is matched to the transmission line (that is, a 50 Ω TDR is used on 50 Ω lines). This is desirable because the reflected pulses arrive at the sending end, are absorbed by the generator impedance, and do not get reflected back toward the load again. An excellent way to realize how little you really know about pulses on transmission lines is to use a TDR that is not matched to the line. In practice this is avoided, as the backward-traveling pulse reflected from the load and forward-traveling pulse reflected

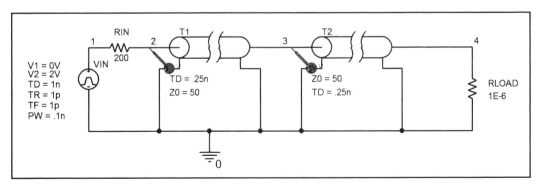

Figure 9.15 — Transmission line schematic with VPULSE input.

from the generator coincide at the input and make interpretation of the display quite difficult.

This last example illustrates a pulsed TDR, with a 200 Ω output impedance, connected to a 50 Ω line. As if that's not bad enough, the load on the line is a short circuit. When the voltage at the input is observed, confusion reigns. In order to make what is happening understandable, we can use *PSpice* to peek at the voltage in the middle of the transmission line.

Figure 9.15 shows the pulse generator with 200 Ω output impedance connected to two 50 Ω transmission lines in series. The characteristics of the input pulse are defined in the schematic design. The pulse voltage source, VIN, will start at 0 V and rise to 2 V after a 1 ns delay. The rise time and fall time will both occur in 1 ps. The width of the pulse will be 0.1 ns. Each line has a time delay of 0.25 ns, so the total one-way transmission time is 0.5 ns. Using two lines in series allows us to view the voltage at the midpoint, where the forward-traveling and reverse-traveling pulses do not coincide. In the schematic design, the load resistor is listed with a 1E-6 Ω value (0.000001 Ω), as *PSpice* does not allow 0 Ω resistors.

In order to observe the reflection of the input pulse, a Transient Analysis will be carried out on the schematic design with a run time starting at 0 ns

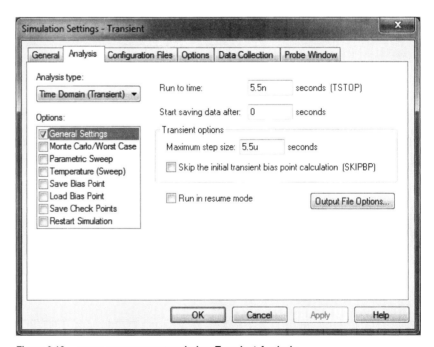

Figure 9.16 — SIMULATION SETTINGS window, Transient Analysis.

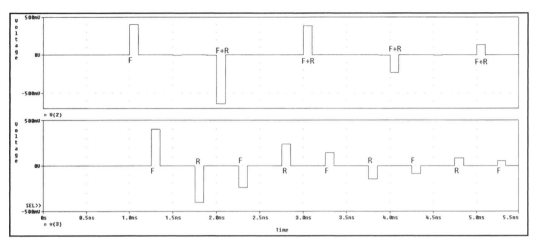

Figure 9.17 — *AMS Simulator* window plotting voltage from both node 2 and node 3 (F, R, and F+R labels have been added to indicate direction of pulse).

to 5.5 ns with a step size of 5.5 µs to provide an adequate resolution to the output plot in the *PSpice* PROBE window. Simulation parameters are shown in **Figure 9.16.**

The PROBE display of analysis results seen **Figure 9.17** shows the voltage at the generator, V(2), on the upper plot and the midpoint voltage, V(3), on the lower plot. Labels have been added in the *PSpice* plots to mark each pulse with its direction of propagation — F (forward) or R (reverse). On the upper plot, the first pulse, of 400 mV, is seen at 1 ns. The 400 mV is easily explained as the voltage division of the 2 V generator pulse between the 200 Ω generator resistance and the 50 Ω line impedance. However, at 2 ns a pulse is seen, whose amplitude of –640 mV may not be so obvious; this is the sum of the –400 mV pulse reflected back from the load and the –240 mV pulse reflected back towards the load by the generator.

The voltage reflection coefficient of the load is –1.0, while the generator has a voltage reflection coefficient of +0.6. Observing the lower plot of the voltage at the transmission line midpoint, you can see much more clearly what is happening. Every time a forward-traveling pulse (labeled F) is reflected at the load, it comes back with the opposite polarity, multiplied by –1; every time a reverse-traveling pulse (labeled R) is reflected at the generator, it comes back with the same polarity, but multiplied by +0.6.

PSpice has been used here to give insight into the behavior of a transmission line circuit, which would be difficult to glean in the laboratory.

In Summation

As you can see, *PSpice* is an excellent tool in the sense that it can simulate and solve transmission line problems in a matter of seconds, whereas carrying out manual calculations can be tedious and time consuming. But it is important to note that *PSpice* recognizes only a lossless transmission line. To compensate for this, if you wish to simulate a more realistic example, it is possible to add distributed losses to a transmission line by breaking the transmission line into a number of short lengths and inserting resistances in series.

Chapter 10

Subcircuits

Many circuits are made with basic electronic building blocks, such as operational amplifiers (op-amps), logic gates and comparators. For example, an active filter might contain six identical op-amps. One way to create an input schematic for such an active filter to be analyzed using the *PSpice* software would be to place the op-amp six times in the schematic capture window. In addition, resistors and capacitors certainly will be attached to those op-amps. This approach can be tedious and monotonous work, especially when you begin to work with larger and more complex schematic designs. Fortunately, *PSpice* permits us to define an often-used or redundant circuit block in a single new component model. This circuit block can then be used many times, like a part, each time the circuit block is used in a schematic design. *PSpice* calls such circuit blocks *subcircuits*.

Linear Dependent Sources

This chapter will demonstrate how to create a subcircuit using the *PSpice* software. We will also take this opportunity to introduce linear dependent source components. *PSpice* lets you use dependent (or controlled) voltage sources and current sources in describing circuits. This can be helpful in making simple models of semiconductor devices, such as BJTs, FETs, and op-amps.

The four kinds of linear dependent sources are shown in **Table 10.1**.

Table 10.1
Linear Dependent Sources

Linear Dependent Source	Abbreviation	PSpice Component Name (see Figure 10.1)
Voltage-controlled voltage source	VCVS	E
Current-controlled current source	CCCS	F
Voltage-controlled current source	VCCS	G
Current-controlled voltage source	CCVS	H

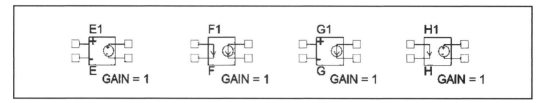

Figure 10.1 — Linear dependent source schematic symbols (see Table 10.1).

The schematic symbols for each of the four linear dependent sources are shown in **Figure 10.1.** Each of these source components consists of four terminals. Two terminals are sensory terminals that detect an input voltage, in the case of the VCVS or VCCS (*voltage-controlled voltage source or voltage-controlled current source*), or input current in the case of the CCCS or CCVS (*current-controlled current source or current-controlled voltage source*). The input to the sensor terminals, in conjunction with a user-defined gain constant, will determine how much voltage or current will be present at the output terminals of the linear dependent source.

Using component E, the voltage-controlled voltage source, a circuit will be constructed in the Cadence *Allegro Design Entry/OrCAD Capture* schematic capture window to show how to create a subcircuit using the *OrCAD/PSpice* software. After the subcircuit has been created, it will be demonstrated that a VCVS exhibits similar behavior to a typical op-amp, in the sense that it is possible for a user to create a certain amount of voltage gain that can be seen between an input and output voltage.

Creating a Subcircuit Block

First, let's reduce this basic schematic design into a subcircuit block (see **Figure 10.2**). In Figure 10.2 the VCVS component has been placed in the *OrCAD* drawing space. The VCVS, as with any of the linear dependent sources, features the sensor terminals on the left of the schematic component

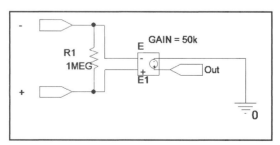

Figure 10.2 — VCVS schematic design to simulate op-amp.

and the output voltage or current to be present within the terminals to the right of the component. Thus, our input pins (which we will talk about in just a moment) have been attached to these. Also, note the presence of R1. This large 1 MΩ resistance between the input terminals is characteristic of the enormous input impedance of a typical op-amp. A linear dependent source exhibits almost ideal behavior, so R1 is there only to help mimic the input terminals of an op-amp model. To the right of the terminals, the negative output has been grounded, and the output has been designated at the positive output pin of the VCVS component. With the negative pin grounded, all the potential developed between the output pins will be seen at the positive terminal/output pin, and again, will simulate a singular output of an op-amp.

To summarize, the input pins (designated + and −) at the left of the schematic are similar to the input pins of an op-amp. An input potential attached to these pins will be sensed and acted upon by the VCVS component and produce an output based on a given value of gain. In the case of this sample circuit, the gain has been specified as 50,000.

Now, to place and designate the input and output pins of a subcircuit, select the button in the right toolbar titled PLACE PORT. You will be presented with the window shown in **Figure 10.3**. Scroll through the

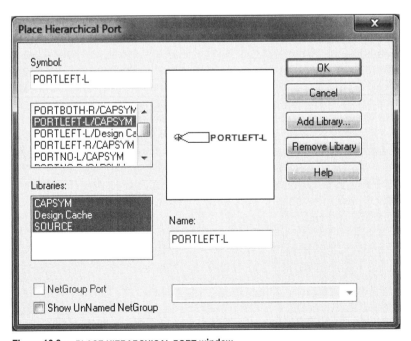

Figure 10.3 — PLACE HIERARCHICAL PORT window.

available ports and find ports PORTLEFT-L and PORTRIGHT-R. Placing the port called PORTLEFT-L will create a terminal that will appear on the left of the subcircuit block that you will create, and placing PORTRIGHT-R will create a terminal that will appear on the right of your subcircuit block. Two PORTLEFT-L pins have been placed on the input side of the design and have been named simply + and –. The terminal – denotes the port that connects directly to the negative sensory terminal of the VCVS, and the terminal + connects directly to the positive sensory terminal of the VCVS. The output terminal of the VCVS component connects directly to a PORTRIGHT-R port and has been named OUT. Thus a single output pin will appear on the soon-to-be-created subcircuit block named OUT.

Now we turn to the first step in creating the subcircuit block based on this design. After drawing the schematic to be reduced into subcircuit form, save the design and select the CREATE NETLIST option found in the PSPICE menu, as shown in **Figure 10.4**. This will present you with a CREATE NETLIST window. Click the PSPICE tab at the top of the window so that the CREATE NETLIST window looks similar to **Figure 10.5**. If this CREATE NETLIST window does not open, you may need to use another method of creating the netlist. There are subtle differences between some of the most recent versions of the software, and whether or not you experience this problem of not being able to open the CREATE NETLIST window depends on

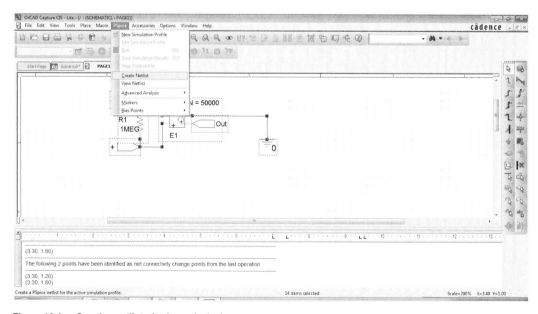

Figure 10.4 — Creating netlist of schematic design.

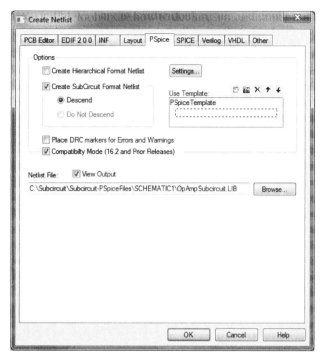

Figure 10.5 — CREATE NETLIST window.

the version of the *OrCAD/PSpice* software you are using. Another process that surely will work for creating the netlist is to select the project tab in the *OrCAD* window. This will display the project folder and all associated files in a hierarchal format. Select the schematic page that contains this design by clicking on it, and then select the CREATE NETLIST button from the toolbar at the top of the screen. For more information regarding this method, refer to Chapter 3.

Once the CREATE NETLIST window has been opened and the PSPICE tab has been selected, check the box to create a subcircuit format netlist with the DESCEND bubble chosen. This means that a netlist will be created that describes the schematic to be contained within the subcircuit. The fact that this subcircuit descends means that once the subcircuit has been implemented in a new design, the subcircuit netlist will lie on a hierarchical tier below the top schematic page of the design. The text box at the bottom of the window designates the destination and the file name for the netlist of the subcircuit block. The VIEW OUTPUT box has been checked in this window, so after pressing OK, the completed subcircuit netlist file will be opened in a new window.

Figure 10.6 — Netlist for VCVS design.

Figure 10.6 shows that the netlist for the subcircuit block based on the schematic design created earlier in the chapter has been saved under the file name "OpampSubcircuit".

Introduction to *Model Editor*

Open the *Model Editor* program included with the *OrCAD/PSpice* software. Select OPEN in the FILE menu and find the netlist file for the subcircuit block you created (see **Figure 10.7**). The file extension for the netlist that you have created is .lib. This denotes a *Model Editor* library file.

Open the netlist and the MODEL EDITOR window will resemble what is shown in **Figure 10.8**. You will see the same netlist text file that was shown in the *OrCAD* window after you generated the netlist earlier. The name of the subcircuit block will be SCHEMATIC1. This is designated in the column at the left side of the screen.

Under the FILE dropdown menu, select the option EXPORT TO CAPTURE PART LIBRARY to open the window shown in **Figure 10.9**. This will create the schematic symbol for the subcircuit block that will appear in the *PSpice* library. The top text box is the destination for the library file that was just opened; the bottom test box designates the library name that will contain this subcircuit block SCHEMATIC1. This file extension is .olb, which stands for *OrCAD* library. The *OrCAD* library name chosen for this example is OpampSubcircuit.

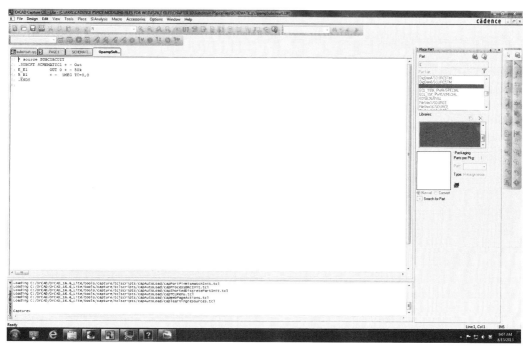

Figure 10.7 — Opening the *Model Editor* library file.

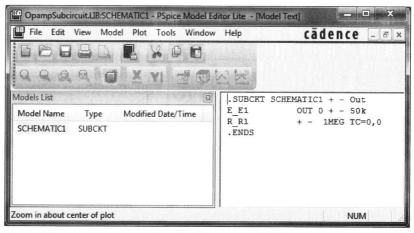

Figure 10.8 — MODEL EDITOR window displaying imported VCVS design netlist.

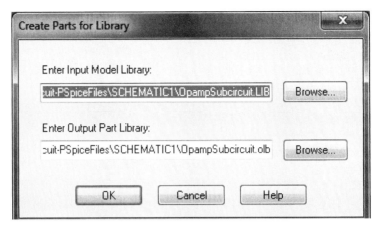

Figure 10.9 — CREATE PARTS FOR LIBRARY window.

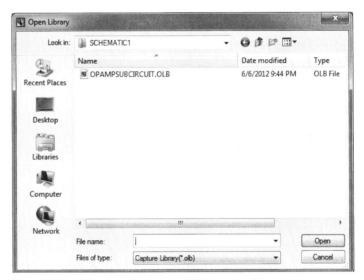

Figure 10.10 — Library containing new VCVS subcircuit.

Create a new project and open a blank schematic page in the *OrCAD* schematic capture window. Next, add a new library in the PLACE PART toolbar by clicking the ADD LIBRARY button previously mentioned in Chapter 2. Browse and find the name of the *OrCAD* library that was just created using the *Model Editor* program. In this case the library name chosen is OpampSubcircuit.olb (see **Figure 10.10**). Open this library.

This library now is active in the schematic window along with all of the other components that you have been using up to this point. There is

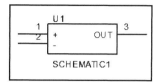

Figure 10.11 — Schematic symbol for VCVS subcircuit.

only one component in this library, and it is shown in **Figure 10.11**. It is the subcircuit block SCHEMATIC1, and it features three pins that correspond with the ports placed in the schematic design in the beginning of the chapter.

Schematic Design Incorporating the Subcircuit Block

Figure 10.12 shows a schematic that was drawn utilizing this new subcircuit model. A 100 µV, 1 kHz sinusoidal input has been applied to the positive terminal of the reduced VCVS circuit while the negative terminal has been grounded. A 1 kΩ load resistor has been attached to the output

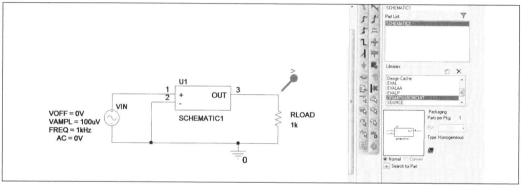

Figure 10.12 — Schematic design featuring VCVS subcircuit.

Figure 10.13 — SIMULATION SETTINGS window, configuring the subcircuit library.

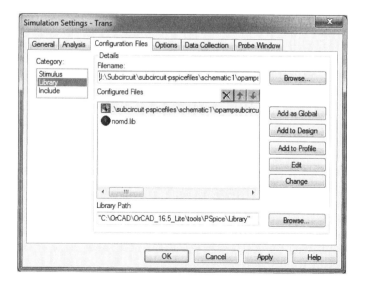

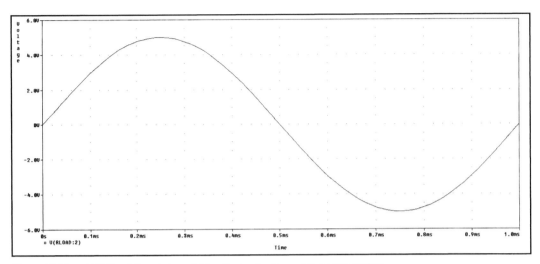

Figure 10.14 — *AMS Simulator* window plotting the load voltage with respect to time.

terminal. Recall that the VCVS component within this subcircuit has a gain of 50,000, so it is expected that we will see an output of 5 V after simulating this circuit.

Before simulating, it is essential to configure the new library. In the SIMULATION SETTINGS window, select the CONFIGURATION FILES tab. Click LIBRARY and then browse to find the OpampSubcircuit.olb file (see **Figure 10.13**). Click ADD TO DESIGN and apply these changes after opening the library file. This action allows the *PSpice* software to attach the underlying netlist data for the SCHEMATIC1 component to the simulation profile. Even though the subcircuit block is visible in the schematic design, the netlist data is not associated with the schematic from a simulation standpoint. If the user neglects to configure the library for the subcircuit component, a simulation will yield errors. Most likely, the *PSpice* software will inform the user that the component is undefined and that all of the attached traces and components are floating.

A Transient Analysis was carried out on this circuit with a start time of 0 s and an end time of 1 ms with a step size of 1 µs. As predicted, the output voltage measures exactly 5 V (see **Figure 10.14**).

In Summation

Being able to create a subcircuit can significantly reduce the complexity of a design and can be especially helpful in an instance where a circuit might feature redundant branches or networks. These subcircuits can be utilized to simplify some of the sample circuits in the next chapter.

Chapter 11

Sample Circuits

This chapter is devoted entirely to sample circuits. Analysis will be carried out on various examples to effectively put into practice all the concepts discussed in previous chapters. The early examples of this chapter will carry out Bias Point Calculations on dc circuits. This will be followed by examples using sweepable dc parameters, then ac parameters, and finally, examples using the Transient Analysis.

These examples were designed with the intention of expanding further on the topics that were previously discussed and may delve even a little deeper into some of the more advanced functions of the Cadence software.

Bias Point Examples

Figure 11.1 shows a dc circuit with one battery and nine resistors. The problem is to find the dc voltage across RG, the 600 Ω resistor. As done in

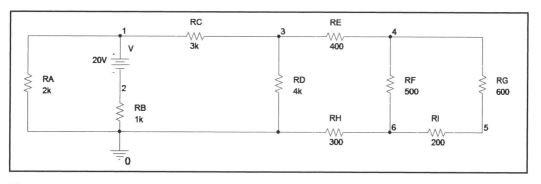

Figure 11.1 — Nine-resistor dc circuit.

previous chapters, we will draw the schematic design and perform a small signal analysis, which will cause the voltage at each node and the voltage source current to be printed in the output file.

The outputs in **Figures 11.2, 11.3**, and **11.4** contain the Small Signal Bias Solution calculations, giving measurements of voltage

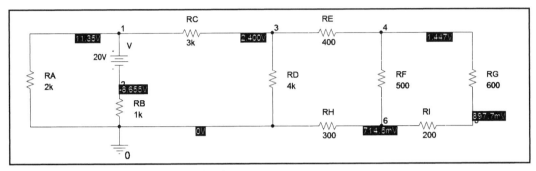

Figure 11.2 — Bias Point results showing only voltages.

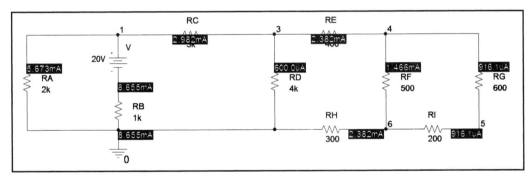

Figure 11.3 — Bias Point results showing only currents.

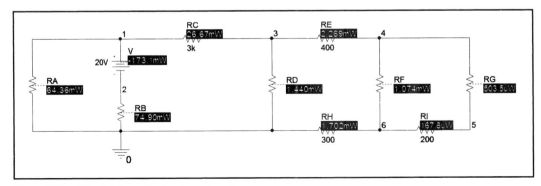

Figure 11.4 — Bias Point results showing only wattages.

with respect to ground, current and power, respectively. Since we want to find the voltage across resistor RG, it will be necessary to subtract manually the voltage at node 5 from the voltage at node 4, as follows:

$$V_{RG} = V(4) - V(5) = 1.4474 - 0.8977 = 0.5497 \text{ V}$$

While it is not necessary for this problem, the Small Signal Bias Solution will also yield the branch currents, and the power dissipated within each component is displayed in the outputs shown Figures 11.3 and 11.4.

The dc circuit in **Figure 11.5** contains a voltage source, a current source, and three resistors. The voltage across the 40 Ω resistor (R2) and the current through the 30 Ω resistor (R1) will be determined. Similar to the previous example, these values can be found using the Small Signal Bias Solution.

Figure 11.6 displays the voltage at each node of the circuit with respect to ground. The voltage measured from node 4 of the circuit is the voltage drop across the 40 Ω resistor. Referring to the output from Figure 11.6, this potential is measured at –5.143 V dc.

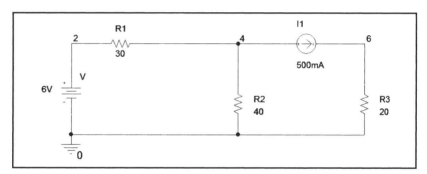

Figure 11.5 — A dc circuit with a voltage source and a current source.

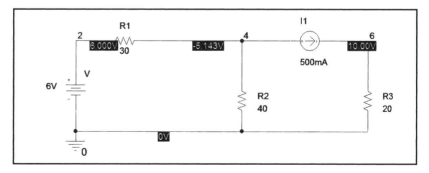

Figure 11.6 — Bias Point results showing only voltages.

Notice in **Figure 11.7** that the current values have been toggled visible in the Cadence *Allegro Design Entry/OrCAD Capture* window. The current has been measured in the branch connecting node 2 to node 4 as 371.4 mA.

The next example will carry out a Bias Point Calculation for a circuit featuring one of the linear dependent sources alluded to in Chapter 10. The schematic diagram in **Figure 11.8** is a dc circuit with a linear voltage-controlled current source, called G. The current through the voltage-controlled current source G is equal to five times V_{RA}, the voltage across the 15 Ω resistor. The problem is to find the voltage across the 6 Ω resistor (RD) and the current through the 25 Ω resistor (RB). In order to easily find the voltage across the 6 Ω resistor, ground (node 0) has been placed at its bottom terminal.

The input schematic drawing shown in Figure 11.8 is simulated

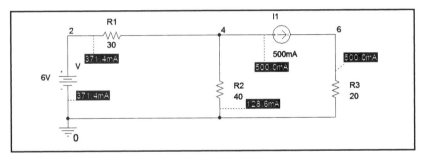

Figure 11.7 — Bias Point results showing only currents.

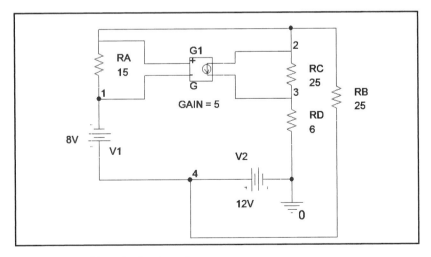

Figure 11.8 — VCCS dc circuit schematic.

using the Small Signal Bias Solution. With reference to the discussion in Chapter 4, just as was seen in that first sample circuit, here the circuit voltages, currents, and wattages can be toggled on and off in the schematic capture window after the Small Signal Bias Solution is complete. **Figure 11.9** displays all the node voltages in the circuit. According to the simulation results, the voltage drop across resistor RD is equal to 1.891 V.

Another element of the problem statement was to find the current across the resistor RB. The circuit currents are displayed in **Figure 11.10**. The Small Signal Bias Solution results indicate that the current across RB, the 25 Ω resistor, is 318.2 mA.

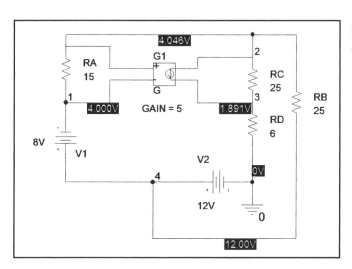

Figure 11.9 — Bias Point results showing only voltages.

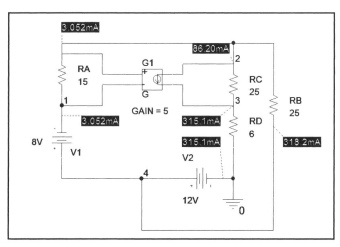

Figure 11.10 — Bias point results showing only currents.

DC Sweep Examples

Figure 11.11 illustrates a test circuit that can generate a volt-ampere curve for a diode. This is a similar method, used previously in Chapter 7, except we will take the simulation further to illustrate other important characteristics of the diode in question. Again, we will be using the 1N4002 diode. A current measurement marker has been placed into the schematic design to measure the current across the diode as the source voltage, VD, sweeps.

The SIMULATION SETTINGS window is shown in **Figure 11.12**. The diode voltage is stepped from 0.6 V to 0.81 V in increments of 2 mV.

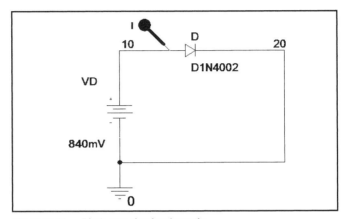

Figure 11.11 — Diode test circuit schematic.

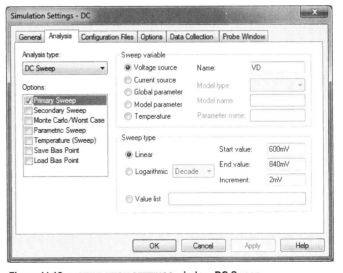

Figure 11.12 — SIMULATION SETTINGS window, DC Sweep.

Figure 11.13 shows the diode volt-amp characteristics. Using the cursor in PROBE, at 600 mV the diode current is 1.763 mA, while at 810 mV the current has risen to 163.763 mA. If the Y axis is made logarithmic rather than linear, a straight line graph results due to the relationship between voltage and current in a diode. Remember that the *PSpice* software has ability to display a trace that plots a mathematical expression. This function is used to the graph the diode's static resistance, VD/ID, in **Figure 11.14**, and

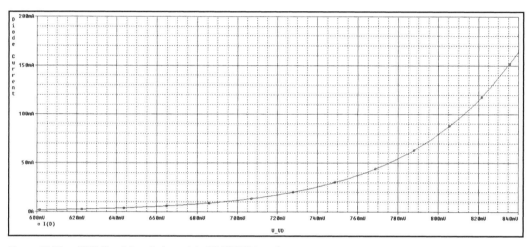

Figure 11.13 — *AMS Simulator* window, plot of D1N4002 transfer curve.

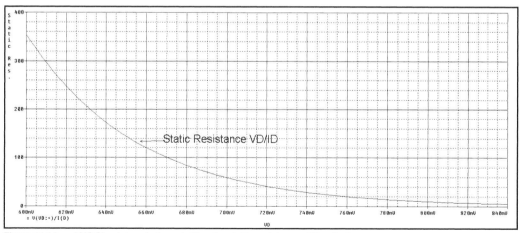

Figure 11.14 — *AMS Simulator* window plotting the static resistance of the diode.

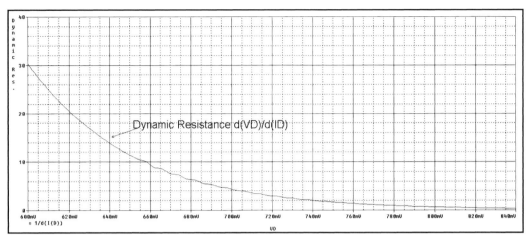

Figure 11.15 — *AMS Simulator* window plotting the dynamic resistance of the diode.

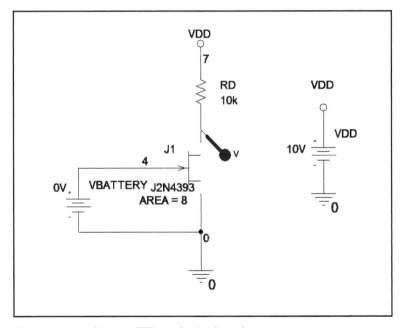

Figure 11.16 — N-Channel JFET test circuit schematic.

the diode's dynamic resistance, d(VD)/d(ID), in **Figure 11.15**. It is interesting to note that at 640 mV, the static resistance is approximately 173 Ω, while the dynamic resistance is much lower, 14 Ω.

Before moving into some real circuits with practical applications, we will carry out the same process used for the 1N4002 diode using a JFET. The process will look very similar to examples carried out in Chapter 8, except we will use other simulation and trace techniques to explore other characteristics of the transistor. The test circuit of **Figure 11.16** uses an N-channel JFET, J1. A graph of drain voltage versus gate-source voltage is desired. The gate-source voltage, VBATTERY, is stepped from –2.5 V to 0.5 V in steps of 20 mV.

Before creating a simulation profile and carrying out a simulation, we will use the Property Editor to modify the physical characteristics of the JFET. Double-click on the JFET to open the PROPERTY EDITOR window (see **Figure 11.17**). Under the AREA column, change the numerical value to "8". However, the element line for the JFET contains an area factor of 8. This means that the physical area of JFET J1 is 8 times larger than the default JFET, and several of the JFET parameters will be affected.

The voltage source, VBATTERY, sweeps from –2.5 V to 0.5 V in increments of 0.2 V. For this particular simulation profile, VBATTERY is the only component with parameters that will be swept. Sweeping the value of this single source will yield a transfer curve for the transistor in the *PSpice* OUTPUT PLOT window. The simulation parameters are defined in the SIMULATION SETTINGS window as shown in **Figure 11.18**.

Figure 11.19 is a graph of drain voltage, V(3), versus gate-source voltage, VBATTERY. The drain voltage begins to drop when VBATTERY

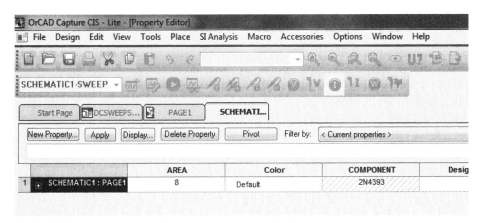

Figure 11.17 — Property Editor for JFET model.

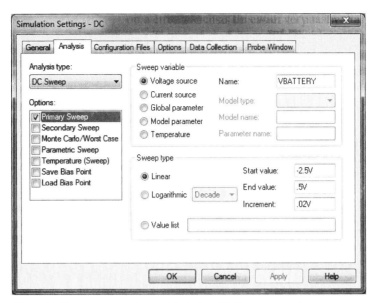

Figure 11.18 — SIMULATION SETTINGS window, DC Sweep parameters.

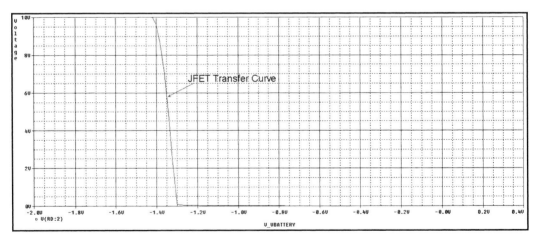

Figure 11.19 — *AMS Simulator* window plotting JFET transfer curve.

reaches –2 V, and the JFET is saturated when VBATTERY is above –1.3 V. The PROBE cursor could be used to get exact numerical data, if needed.

Figure 11.20 is a schematic diagram of a transistor-transistor logic (TTL) 7404 inverter. The output is connected to a 2 kΩ pull-up resistor. In order to determine the transfer characteristic of this circuit, a dc source, VIN, will be swept. The range of the input voltage sweep will cause changes in the output voltage characteristic of a logical 1 changing to a logical 0 at room temperature. The sweep parameters will be discussed below.

To simulate the change in logic level from 1 to 0, the dc voltage source will be swept from 1.10 V to 1.60 V. In order to provide a high-resolution output plot, the DC Sweep will occur in 1 mV incremental steps. These simulation parameters have been outlined in **Figure 11.21**.

The graph of **Figure 11.22** shows that the output voltage stays at nearly 5 V until the input voltage increases to about 1.28 V. As the input voltage rises from 1.28 V to 1.38 V, the output monotonically decreases. For input voltages above 1.38 V, the output is less than 30 mV, indicating that transistor Q4 is well into saturation.

One note of caution: In a circuit that is bi-stable or has hysteresis, such as a Schmitt trigger (a TTL 7414 IC is one example), the use of dc analysis can be problematic. In the dc analysis, previous circuit values are not taken

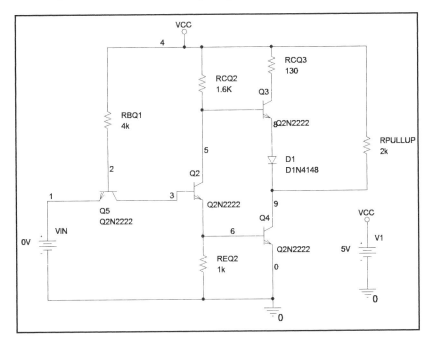

Figure 11.20 — TTL inverter schematic.

into consideration when calculating circuit conditions for the next input voltage. Since there are two possible stable output voltages for a given input voltage (depending on whether the input voltage is rising or falling), *PSpice* may fail to converge or may give incorrect results. A better way to analyze a bi-stable circuit is to use a piecewise linear (PWL) source with a triangular shape and perform a Transient Analysis. When a Transient Analysis is done, *PSpice* does "remember" the previous circuit conditions when calculating at the next time increment. Sample circuits will be used later in this chapter to illustrate the usage and application of a piecewise linear source.

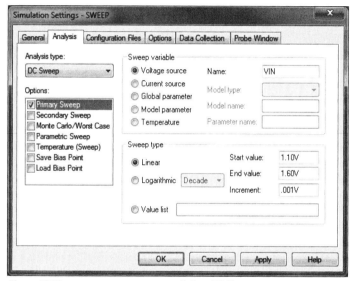

Figure 11.21 — SIMULATION SETTINGS window, DC Sweep parameters.

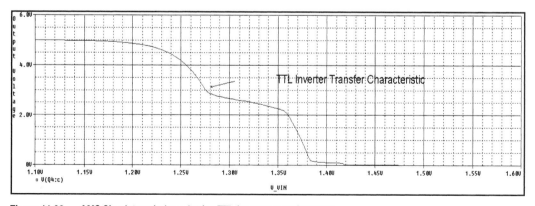

Figure 11.22 — *AMS Simulator* window plotting TTL inverter transfer curve.

AC Sweep Examples

The RC circuit of **Figure 11.23** is connected to a 1 V sinusoidal source. It should have a half-power, or –6 dB, frequency of 500 Hz. A *PSpice* AC Analysis will be used to generate a Bode plot (a graph of the magnitude and phase of the output voltage versus frequency) of the circuit response.

The schematic design of Figure 11.23 calls for the source voltage, V1, to be stepped from 5 Hz to 0.5 MHz logarithmically, with 100 steps per decade of frequency. The output voltage (at node 5) will be plotted in decibels in the OUTPUT PLOT window. These simulation parameters are shown in **Figure 11.24**.

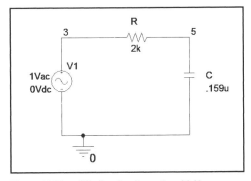

Figure 11.23 — RC circuit schematic with Vac source.

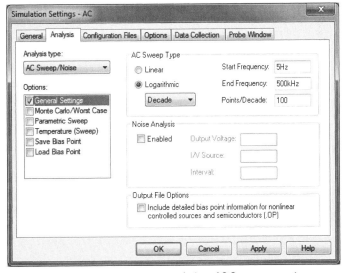

Figure 11.24 — SIMULATION SETTINGS window, AC Sweep parameters.

Figure 11.25 shows a graph of VDB(5) that indicates that the output voltage magnitude is essentially 0 dB at 5 Hz, falls to –3 dB at 500 Hz, and then linearly (on a logarithmic frequency axis) rolls off at 20 dB/decade. The PROBE cursor shows that at 5 kHz the output is –20.035 dB, and at 50 kHz (one decade of frequency above 5 kHz) the output is –39.992 dB.

The phase of the output voltage starts at 0°, is –45° at 500 Hz, and asymptotically approaches –90° as frequency increases above 1000 Hz.

The Wien bridge network is often used to determine the frequency of a low-frequency RC oscillator. The Wien bridge, developed by Max Wien, is a kind of bridge circuit that comprises four resistors and two capacitors. **Figure 11.26** shows half of a Wien bridge, which is the interesting and more difficult to understand half (the other half is the resistors in series). In order to see the response of the network with the output at node 6, the input voltage will be swept logarithmically from 10 Hz to 1000 Hz with 50 frequencies per decade. The schematic input file in Figure 11.26 sets the source voltage magnitude to 1 V ac. Notice that the two resistors are the same value, as are the two capacitors, although they are specified differently.

The screen capture in **Figure 11.27** is subdivided into two plots. The top plot is of the output voltage with respect to the change in frequency, and the bottom plot displays a change in the phase angle with respect to the change in frequency. As shown in Figure 11.27, the phase becomes 0° at 100 Hz. The phase angle becomes positive below and negative above that 100 Hz center frequency. The Wien circuit produces 0° of phase shift at only one frequency, which is also the frequency where the output amplitude is at its maximum value.

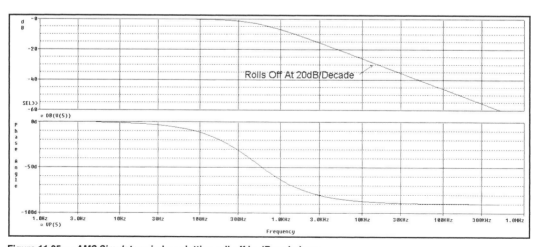

Figure 11.25 — *AMS Simulator* window plotting roll-off in dB and phase.

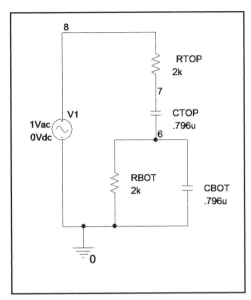

Figure 11.26 — Half Wien bridge circuit schematic.

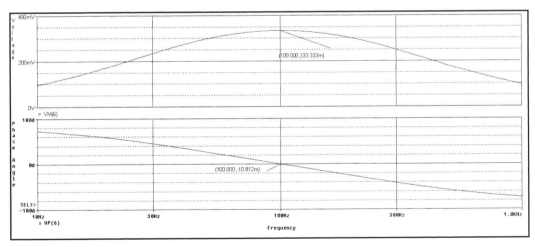

Figure 11.27 — *AMS Simulator* window plotting both voltage magnitude and phase at node 6.

Since at 100 Hz the input voltage is 1 V at 0°, and the output voltage is 0.3338 V at 0°, the gain of the circuit is 1/3, 0°. In order to make an oscillator using this circuit, a non-inverting gain (angle 0°) of 3 (1/(1/3)) would have to be provided. This is true only if the resistors are equal and the capacitors are equal.

The circuit in **Figure 11.28** would be troublesome at best to analyze

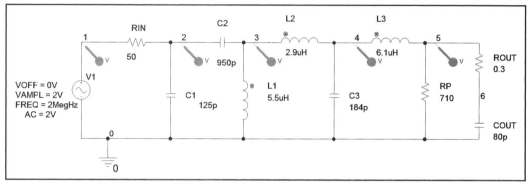

Figure 11.28 — Matching network circuit schematic.

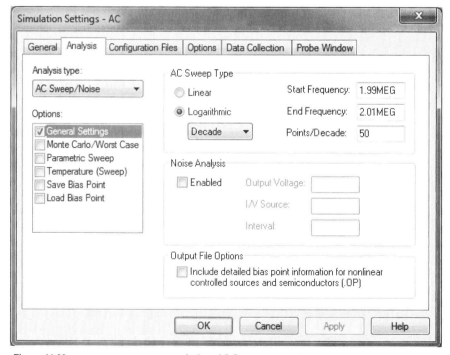

Figure 11.29 — SIMULATION SETTINGS window, AC Sweep parameters.

by hand at a single frequency. It is a matching network that is to be used at several frequencies. However, ROUT and COUT, the resistive and reactive parts of the load impedance, vary with frequency. For this reason the analysis will be at 2 MHz only.

The input file of Figure 11.28 calls for an AC Analysis at one frequency, with six voltages to be printed. The load resistance, ROUT, is between nodes 5 and 6, and the voltage across ROUT and COUT is between node 6 and ground. In order to find the magnitude of the ac voltages at each node of the circuit at one specific frequency, an AC Sweep simulation profile will be created with a relatively small range. These simulation parameters are shown in **Figure 11.29**.

Note that since you need only to observe the behavior of the circuit at 2 MHz, the range of the frequency has been set to start at 1.99 MHz and end at 2.01 MHz, with 50 data points per decade. This will provide you with an output plot displaying only this miniscule portion of the frequency spectrum, with traces centered around 2 MHz.

Voltage measurement markers were placed into the schematic diagram shown in Figure 11.28 at each node of the circuit. The magnitude of each trace has been labeled on the plot shown in **Figure 11.30**.

Obviously, this is not a high-efficiency matching network. Some of the voltages, across reactances VM (4) and VM (6) for example, exceed the input voltage; this is indicative of some resonance effects occurring in the matching network.

One of the limitations of *PSpice* is shown here, which is that a load

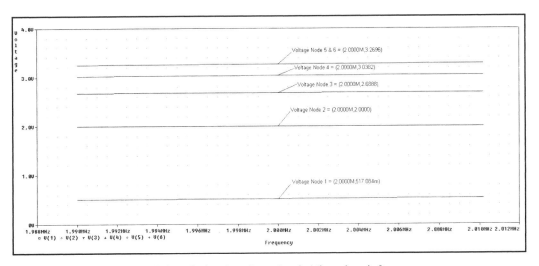

Figure 11.30 — *AMS Simulator* window, plotting the voltage at node 1 through node 6.

that is frequency-dependent cannot easily be described to *PSpice*. Antenna impedances are one common example of this type of load. Since the load impedance is a function of frequency, the input file must be edited (for load impedance) and a *PSpice* analysis must be done separately at each frequency.

It is sometimes useful to be able to compare graphically the performance of two circuits. In this example, the frequency response of two series resonant circuits that are identical except for the loss resistance in each will be compared. The Cadence *Allegro AMS Simulator/PSpice A/D* window will be used to plot the results on one frequency axis in order to make the comparison easy.

Figure 11.31 illustrates the two circuits, which are connected in parallel across a voltage source so a single plot of two response curves can be made.

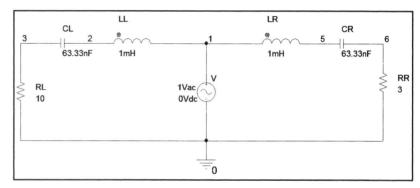

Figure 11.31 — Parallel RLC circuit schematic.

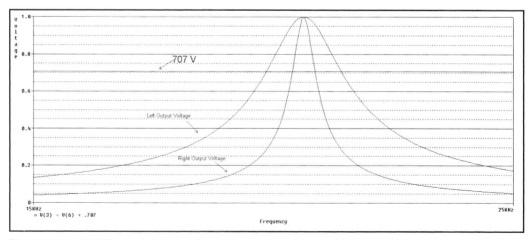

Figure 11.32 — *AMS Simulator* window plotting voltage measured from each RLC branch.

A similar technique can be applied by placing the two circuits in series when a current source is the input.

Each circuit is resonant at 20 kHz; they differ in the series loss resistance, which determines quality factor Q and bandwidth. The source, V, will be swept linearly from 15 kHz to 25 kHz, and the voltage across the two load resistors will be plotted.

Figure 11.32 shows the graph of resistor voltage versus frequency for the two circuits. The substantial difference in the bandwidth of the two circuits is apparent — the left circuit, with a resistance of 10 Ω, has a lower Q and larger bandwidth than the right circuit, which has a resistance of 3 Ω. The PROBE cursors, on the V(3) trace, are located at the half-power frequencies (19.207 kHz and 20.813 kHz), with the bandwidth computed to be 1.5967 kHz.

The analysis of a circuit containing an RF transformer with a coefficient of coupling less than 1 is somewhat tedious, even at a single frequency. To get a frequency response plot is not a task anyone would enjoy (or have time to do) by hand. Figure 11.31 illustrates such a circuit, which is fed by a current source.

One limitation of *PSpice* is apparent in this circuit — each node must have a dc path to ground so that a Small Signal Bias Solution can be found. This is true for any circuit, even if the Bias Solution is meaningless because there are no dc sources in the circuit. Resistor RDCPATH is added to the circuit so that each node will have a dc path to ground. As shown in the schematic in **Figure 11.33**, RDCPATH has a value of 1×10^{12} Ω, so its effect on the circuit performance is negligible. Both primary and secondary windings of the transformer are resonant at 2 MHz, and the coefficient of coupling is greater than optimum coupling. This means that the transformer is overcoupled and will have a very undesirable frequency response.

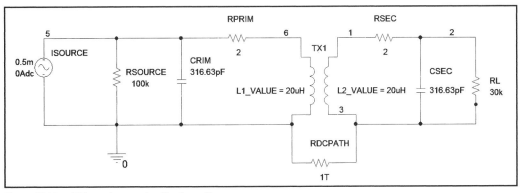

Figure 11.33 — RF transformer circuit schematic.

Figure 11.34, the output plot, shows the classic double-peaked response, where the output had peaks at two frequencies, neither of which is the resonant frequency of 5 MHz. This is a good illustration of the concept of reflected impedance, in which the reactance of the secondary is reflected into the primary circuit and detunes the primary.

A circuit with two ac sources is shown in **Figure 11.35**. One is a voltage source; the other is a current source. The phase voltage at node 5 is the unknown, over the frequency range of 1 kHz to 2 kHz. The circuit requires a dummy to provide a dc path to ground for node 9. In the Small Signal Bias Solution, the capacitor and the ac current source are both open circuits. RDUMMY has a value of 1 TΩ (1×10^{12} Ω), which should have little effect

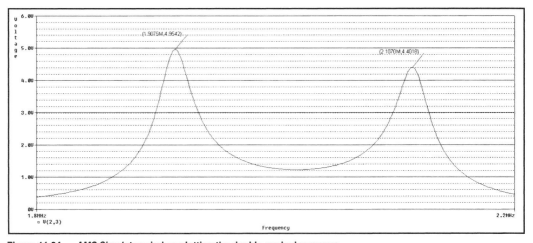

Figure 11.34 — *AMS Simulator* window plotting the double-peaked response.

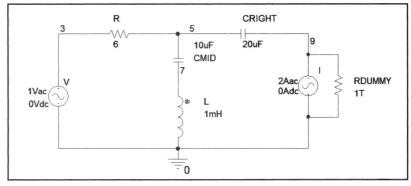

Figure 11.35 — Circuit with two Vac sources.

on circuit behavior. The simulation profile dictates that an analysis occur at 101 frequencies linearly spaced between 1 kHz and 2 kHz. Results of the AC Analysis will be plotted in the *AMS Simulator* window.

In this particular example, the default phase shift of the current source will be –30°. To assign this default parameter, double-click on the current source, I, to open the Property Editor (see **Figure 11.36**). Under the phase column, enter a value of –30°.

Figure 11.37 is a graph of the phasor output voltage. The upper plot indicates that the voltage at node 5 is at a minimum around 1.6 kHz, and that the phase changes abruptly at the same frequency. *PSpice* always expresses phase angles such that the magnitude of the phase angle is 180° or less. Thus, what appears to be an abrupt change in phase angle may simply be due to how it is expressed by *PSpice*.

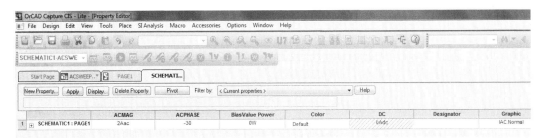

Figure 11.36 — Property Editor for Vac component.

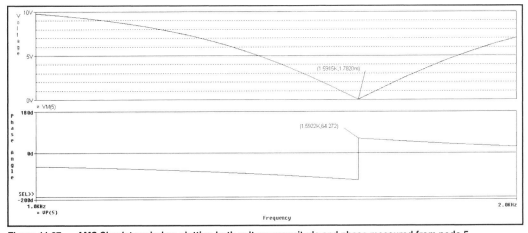

Figure 11.37 — *AMS Simulator* window plotting both voltage magnitude and phase measured from node 5.

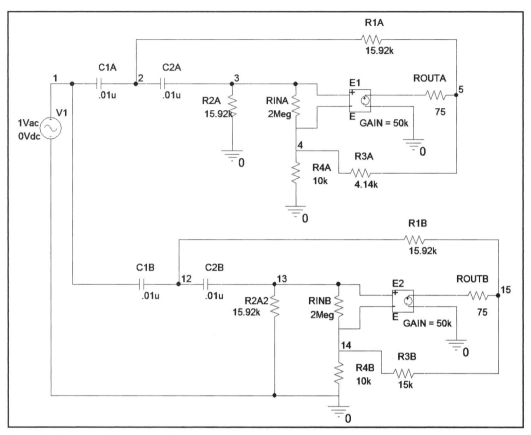

Figure 11.38 — Active filter circuit schematic.

Figure 11.38 shows two active filter circuits connected in parallel across an ac voltage source. Each circuit is a second order high-pass filter; the first has Butterworth response (Butterworth filters are signal processing filters designed to have as flat a frequency response as possible in the passband) and the second has Chebyshev response. (Chebyshev filters are analog or digital filters having a steeper roll-off and more passband ripple, type I, or stop-band ripple, type II, than Butterworth filters.) The response is determined by the closed-loop gain of the op-amp. When constructing the schematic design in the *OrCAD* schematic capture window you may want to create a VCVS subcircuit using the method discussed in the previous chapter to give yourself extra practice and to help simplify the design. You may also choose to simply redraw the circuit with the VCVS linear dependent source.

The input schematic design might have been somewhat simpler if the subcircuit had been used, but it must be noted that you can obtain similar

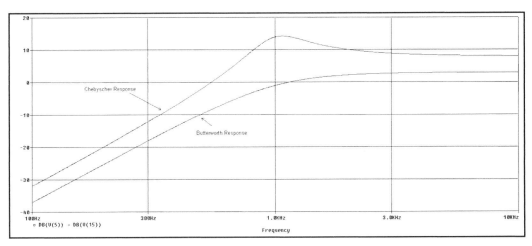

Figure 11.39 — *AMS Simulator* window plotting the frequency response of each filter.

results if you create a similar design using one of the many op-amp models in the *PSpice* library. According to the simulation parameters of this particular example, the frequency of the source is swept logarithmically from 100 Hz to 10 kHz, with 50 frequencies per decade.

The plot of the filters' output, **Figure 11.39**, indicates that the gain of the Chebyshev-response filter is higher than the Butterworth-response filter in all frequencies, with a noticeable peak at 1.122 kHz. While quantitative data could be obtained from this graph by using the CURSOR feature of the *AMS Simulator* window, the graph is of great use in qualitatively understanding the difference between these two filters.

A two-transistor BJT differential amplifier is shown in **Figure 11.40**. It is fed by a voltage source at the base of Q1. This schematic design includes two NPN BJTs. In order to perform an analysis, a simulation profile can be set up so the frequency is swept logarithmically from 100 Hz to 1 GHz.

A voltage marker has been placed on node 6 of the schematic design to produce a voltage trace in the output *AMS Simulator* window. This voltage trace will be modified in the *AMS Simulator* window so the output voltage (which is the gain, since the input voltage magnitude is 1) will be plotted in decibels.

This is a unique sample circuit in the sense that it utilizes components from the Breakout library that don't necessarily have defined characteristics and can be modified. Notice the transistor models in the schematic design are both the QBREAKN component from the Breakout library. The Breakout library (search for this in the *PSpice* library) contains general transistor and

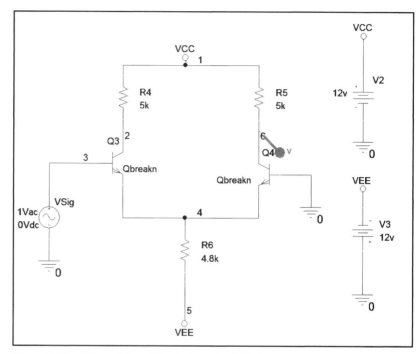

Figure 11.40 — Differential amplifier design using BJTs from the Breakout library.

diode models that are fully customizable to a user's specifications. This model is of an NPN transistor. For this example, we desire a forward beta of 60 and significant junction capacitances representative of discrete devices, not transistors on an integrated circuit. In order to change the characteristics of this transistor model, right-click on either of the QBREAKN transistors and select EDIT PSPICE MODEL. This will open the *PSpice Model Editor* (see **Figure 11.41**).

Model Editor will indirectly amend the element line for the transistor in the circuit's netlist by adding user-specified data. In the center of the MODEL EDITOR window is a large textbox where you will first see a single line of text that reads as follows:

.model Qbreakn NPN

Click inside the text box and modify the line so it reads, as shown in **Figure 11.42**:

.model Qbreakn NPN (BF=60 CJC=16p CJE=30p)

We will discuss the contents of the parentheses in a moment.

The first portion of this line, .model Qbreak NPN, declares that the transistor model that will be modified is the NPN, Qbreakn. When *Model*

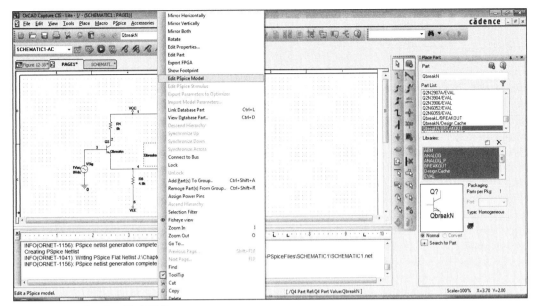

Figure 11.41 — Opening *Model Editor*.

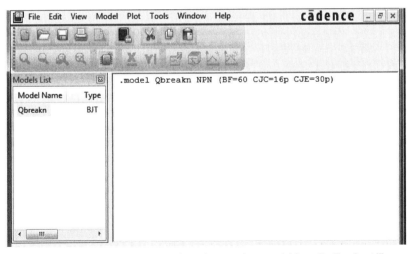

Figure 11.42 — MODEL EDITOR window featuring transistor model from the Breakout library.

Editor first opened, this was the only text associated with the component. The data in the parentheses, that was just added, show the user-specified transistor characteristics. BF stands for beta factor and has been set at a value of 60 for this example. Both CJC and CJE are junction capacitances and have been declared as 16 pF and 30 pF, respectively. These values of capacitance simulate the characteristics of a non-ideal transistor and help dictate the frequency response, specifically rolloff, of the transistor amplifier across the range of the AC Sweep.

There are numerous other customizable parameters for transistors and diodes that can be altered in *Model Editor*; however, there are so many more that it would require a separate text to provide an adequate definition of them all. For more information regarding these parameters, a complete list is available through Cadence *PSpice*.

Here's one last note in regard to *Model Editor*: If a user is to enter enough parameters to take up more than one line of text in the MODEL EDITOR window, a "+" character must be added at the beginning of each line to include any data that may have wrapped to this next line. For example, assume in this example that the junction capacitances wrapped to the next line. Then the MODEL EDITOR window would read as shown below:

.model Qbreakn NPN (BF=60

+ CJC=16p CJE=30p)

It is important to understand that the AC Analysis uses small signal models for all devices. The input voltage of 1 V is hardly a small-signal input, and it causes a great deal more than a small-signal output. Take a look at **Figure 11.43**, which shows an output voltage in the passband of

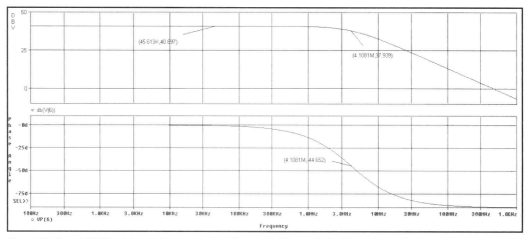

Figure 11.43 — *AMS Simulator* window measuring phase and dB at node 6 with respect to frequency.

40.897 dBV, or 111 V. This is a ridiculous result and could not happen in the actual circuit because the power supplies would limit the output to about 12 Vpp. However, it is convenient to use 1 V as the input magnitude, since the output voltage is then the same as the circuit gain.

If the input were specified as 1 kV, the output would simply be 111 kV. Clearly, you must be careful when interpreting results of an AC Analysis to be sure that the numbers are reasonable. If you wanted to see the effects of nonlinearities, such as clipping of the output due to the transistors becoming cut off and saturated, a Transient Analysis would have to be performed. The Transient Analysis does not use a small-signal model as does the AC Analysis.

The Bode plot shows the half-power frequency to be at 4.1 MHz, as indicated by the 3 dB (2.958 dB is shown) drop in VDB(6) and the phase angle of –45° near that frequency.

Transient Analysis Examples

In **Figure 11.44**, a capacitor and resistor in series are connected to a triangle voltage source. The triangular waveform is made using the PWL function and varies between 0 V and 1 V, with a period of 2 ms.

After the schematic design is drawn into the *OrCAD* schematic capture window, the characteristics of the piecewise function need to be defined. To do this, double click on the VPWL voltage source, VIN, to open the Property Editor for this component. If they are not already present, you must create columns for each timing increment of the triangular waveform. For a triangular waveform to run for a total of 6 ms, it will require seven

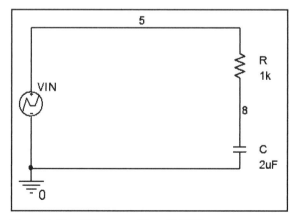

Figure 11.44 — RC low-pass filter with VPWL source.

increments of time, ranging from 0 ms to 6 ms. Simply click the NEW COLUMN button and add columns as shown in **Figure 11.45**. In column T1 enter a time of 0 ms, and for each following column increment this value of time by 1 ms.

Next, voltage magnitudes need to be defined at each increment of

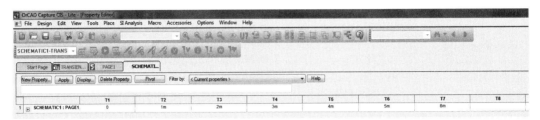

Figure 11.45 — Property Editor for VPWL displaying timing instances.

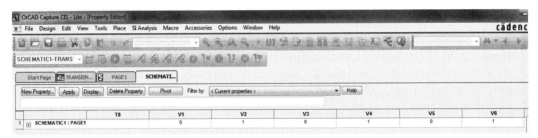

Figure 11.46 — Property Editor for VPWL displaying voltages associated with each timing instance.

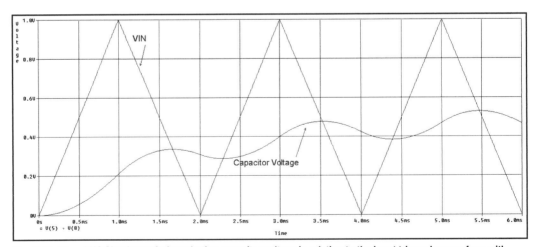

Figure 11.47 — *AMS Simulator* window plotting capacitor voltage in relation to the input triangular waveform with respect to time.

time. Create voltage columns for each of the seven voltages, similar to what is shown in **Figure 11.46**. When the Transient Analysis runs, the PWL waveform will be generated based on each pairing of voltage magnitude and time; T1 will be associated with the magnitude of V1, T2 with V2, all the way through T7 and V7.

Figure 11.47 features this triangle voltage from 0 ms to 6 ms only. Note that there are seven time-voltage pairs in the PWL definition (see Figures 11.45 and 11.46). A simulation profile will be created for a Transient Analysis that ranges from 0 ms to 6 ms, using time steps of 0.01 ms.

An examination of the graph of the capacitor voltage in the output file, shown in Figure 11.47, indicates that the capacitor voltage started at 0 V and had nearly reached steady-state in 6 ms. The average value of the triangle input voltage is 0.5 V.

For analyses that require many cycles of a periodic triangle waveform, using a PWL source can become rather tedious. A better way to generate a periodic triangle, pulse, saw tooth, etc. waveform is to use the PULSE function (the VPULSE component in the *PSpice* library). When doing so, it is important to use a tiny (compared to the period) but non-zero value of pulse width. The rise time and fall time are set to equal half the period of the triangle waveform.

The PWL source is a unique source as it gives the user the opportunity to deliver any conceivable periodic or dynamic voltage/current to a circuit. It is a broad concept, but might be better understood from the standpoint of the component's element line within the netlist. An element line will lay out all the instances of time and voltage/current in a single line of data. For example, a periodic triangle waveform identical to the PWL triangle in this example could be produced by the following element line:

VIN 5 0 PWL(0 0 1m 1 2m 0 3m 1 4m 0 5m 1 6m 0)

Outside of the parentheses, PWL designates that this is a PWL voltage source. Inside the parentheses, the data can be broken up into pairs. The first two values are both 0; the first value of the pair is time, and the second is the magnitude of voltage. In a sense, this can be looked at as similar to an ordered pair. What this means is at the time of 0 s, the source will output 0 V. Moving on to the next pair, the values are 1m 1. This means that at 1 ms the source will output 1 V. The fact that this is a PWL source means that between the times declared there will be a linear rise or fall between the declared magnitudes of voltage. This example of a triangular waveform is ideal for this type of source. Based on the values discussed in the first two pairs, it is expected that we will see a linear rise from starting from 0 V at 0 s to 1 V at 1 ms depicting the rising edge of a triangular waveform.

To demonstrate an alternative method of generating a triangular waveform, a VPULSE source could also be used. Its element line would look much different, as shown below.

VIN 5 0 PULSE(0 1 0 1m 1m 1n 2m)

Reading from left to right in the parentheses, we see the initial pulse value is 0 V, the pulsed value is 1 V, the delay time is 0, the rise time is 1 ms, the fall time is 1 ms, the pulse width is 1 ns (one millionth the rise or fall time), and the period is 2 ms. After 1000 cycles or so, the 1 ns pulse width, which should be 0, will start to introduce a slight error (a cumulative lengthening of the period by 0.0001% each cycle). However, the error is so slight that it doesn't much matter for any practical purpose.

Using either the VPWL or the VPULSE source, *PSpice* will produce virtually the same output plot.

In **Figure 11.48**, an inductor and capacitor in parallel (tank circuit) are connected to a pulse current force. The tank circuit has a resonant frequency of 1007 Hz. This is a circuit that can be realized on paper only, as it is completely lossless. The pulse is very short, with a magnitude of 1 A.

As in the previous example, the user must define the characteristics of the current PWL component, IPWL, using the Property Editor. In this example, the only difference is that each pair will be comprised of an increment of time and a magnitude of current instead of voltage. **Figure 11.49** lists all important values of time, and **Figure 11.50** lists the magnitudes of the source current that will correspond with those times. **Figure 11.51** depicts an output plot measuring the current pulse from the PWL current source, I-IN. **Figure 11.52** plots a measurement of the tank voltage. The pulse starts at 1 ms, and the tank voltage begins ringing at the same time. After the pulse

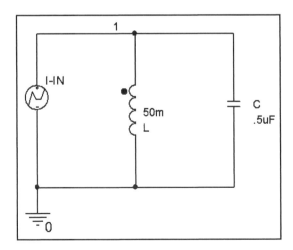

Figure 11.48 — LC circuit with IPWL source.

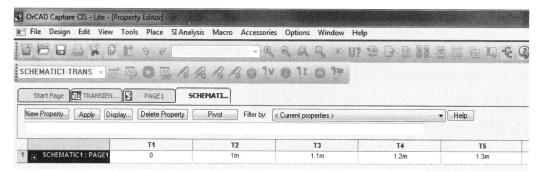

Figure 11.49 — Property Editor for IPWL displaying timing instances.

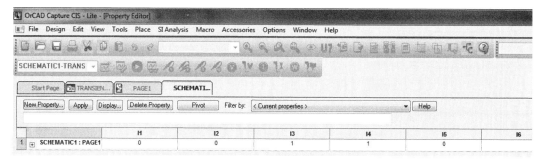

Figure 11.50 — Property Editor for VPWL displaying current magnitudes associated with each timing instance.

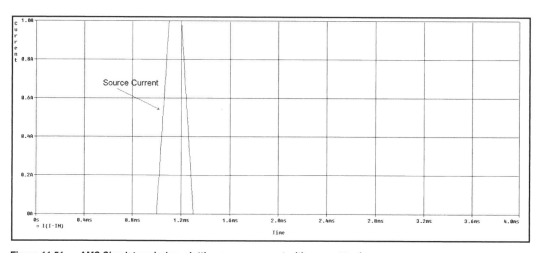

Figure 11.51 — *AMS Simulator* window plotting source current with respect to time.

ends at 1.3 ms, the tank voltage continues to ring with constant amplitude. This is due to the total absence of loss in the circuit; the tank will ring for as long as you care to do the Transient Analysis.

A PWL source can be used to make any arbitrary waveform whatsoever, providing you have the patience to put all the information on the element line. This example contains an electrocardiogram (EKG) voltage over a time interval of 500 ms, with data points every 5 ms.

The circuit shown in **Figure 11.53** is a single-pole RC low-pass filter connected to the EKG voltage. It has a breakpoint frequency of 10 Hz.

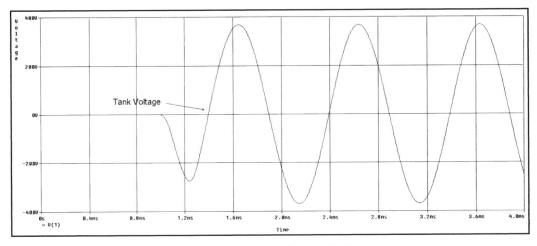

Figure 11.52 — *AMS Simulator* window plotting tank voltage with respect to time.

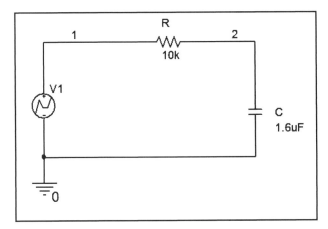

Figure 11.53 — RC low-pass circuit with VPWL (EKG) source.

After entering the low-pass filter circuit into the *OrCAD* schematic capture workspace, one must enter the timing and voltage pairs of the PWL voltage source, V1. Using the same method as shown in the two previous examples, adding new columns into the Property Editor, the voltage and timing pairs are incorporated into the schematic design to replicate a beating heart. You may recall how tedious it was when entering the timing/magnitude pairs in the previous examples, and a waveform as complex as this can be rather monotonous and time consuming, but it can be done (see **Figure 11.54**). However, there is a much easier method.

With the PWL element line discussed earlier in this chapter, it is possible to simulate a circuit using only a netlist text file. To refresh yourself on this process, refer to the second netlist discussion (Chapter 5). Using a VPWL source named "VGEN", a netlist was created in a Text Editor as shown in

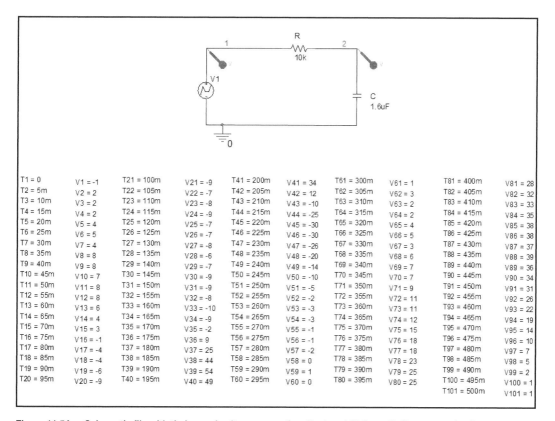

T1 = 0	V1 = -1	T21 = 100m	V21 = -9	T41 = 200m	V41 = 34	T61 = 300m	V61 = 1	T81 = 400m	V81 = 28
T2 = 5m	V2 = 2	T22 = 105m	V22 = -7	T42 = 205m	V42 = 12	T62 = 305m	V62 = 3	T82 = 405m	V82 = 32
T3 = 10m	V3 = 2	T23 = 110m	V23 = -8	T43 = 210m	V43 = -10	T63 = 310m	V63 = 2	T83 = 410m	V83 = 33
T4 = 15m	V4 = 2	T24 = 115m	V24 = -9	T44 = 215m	V44 = -25	T64 = 315m	V64 = 2	T84 = 415m	V84 = 35
T5 = 20m	V5 = 4	T25 = 120m	V25 = -7	T45 = 220m	V45 = -30	T65 = 320m	V65 = 4	T85 = 420m	V85 = 38
T6 = 25m	V6 = 5	T26 = 125m	V26 = -7	T46 = 225m	V46 = -30	T66 = 325m	V66 = 5	T86 = 425m	V86 = 38
T7 = 30m	V7 = 4	T27 = 130m	V27 = -8	T47 = 230m	V47 = -26	T67 = 330m	V67 = 3	T87 = 430m	V87 = 37
T8 = 35m	V8 = 8	T28 = 135m	V28 = -6	T48 = 235m	V48 = -20	T68 = 335m	V68 = 6	T88 = 435m	V88 = 39
T9 = 40m	V9 = 8	T29 = 140m	V29 = -7	T49 = 240m	V49 = -14	T69 = 340m	V69 = 7	T89 = 440m	V89 = 36
T10 = 45m	V10 = 7	T30 = 145m	V30 = -9	T50 = 245m	V50 = -10	T70 = 345m	V70 = 7	T90 = 445m	V90 = 34
T11 = 50m	V11 = 8	T31 = 150m	V31 = -9	T51 = 250m	V51 = -5	T71 = 350m	V71 = 9	T91 = 450m	V91 = 31
T12 = 55m	V12 = 8	T32 = 155m	V32 = -8	T52 = 255m	V52 = -2	T72 = 355m	V72 = 11	T92 = 455m	V92 = 26
T13 = 60m	V13 = 6	T33 = 160m	V33 = -10	T53 = 260m	V53 = -3	T73 = 360m	V73 = 11	T93 = 460m	V93 = 22
T14 = 65m	V14 = 4	T34 = 165m	V34 = -9	T54 = 265m	V54 = -3	T74 = 365m	V74 = 12	T94 = 465m	V94 = 19
T15 = 70m	V15 = 3	T35 = 170m	V35 = -2	T55 = 270m	V55 = -1	T75 = 370m	V75 = 15	T95 = 470m	V95 = 14
T16 = 75m	V16 = -1	T36 = 175m	V36 = 9	T56 = 275m	V56 = -1	T76 = 375m	V76 = 18	T96 = 475m	V96 = 10
T17 = 80m	V17 = -4	T37 = 180m	V37 = 25	T57 = 280m	V57 = -2	T77 = 380m	V77 = 18	T97 = 480m	V97 = 7
T18 = 85m	V18 = -4	T38 = 185m	V38 = 44	T58 = 285m	V58 = 0	T78 = 385m	V78 = 23	T98 = 485m	V98 = 5
T19 = 90m	V19 = -6	T39 = 190m	V39 = 54	T59 = 290m	V59 = 1	T79 = 390m	V79 = 25	T99 = 490m	V99 = 2
T20 = 95m	V20 = -9	T40 = 195m	V40 = 49	T60 = 295m	V60 = 0	T80 = 395m	V80 = 25	T100 = 495m	V100 = 1
								T101 = 500m	V101 = 1

Figure 11.54 — Schematic file with timing and voltage properties displayed (Schematic Capture method).

Figure 11.55. Instead of undergoing the long process creating and naming new columns for timing and voltage in the Property Editor, all of the data was simply typed into the text file and simulated using the *AMS Simulator*.

Either way you choose to run this simulation, the software will yield the same output as shown in **Figure 11.56**. Referring to Figure 11.55, the PWL independent source element "line" actually requires 16 lines, using the

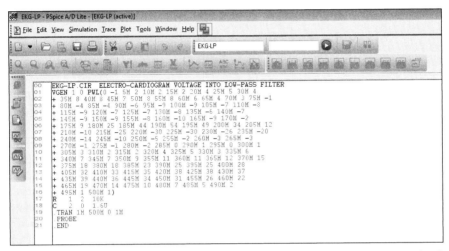

Figure 11.55 — *AMS Simulator* window with an active netlist depicting the schematic design (Netlist method).

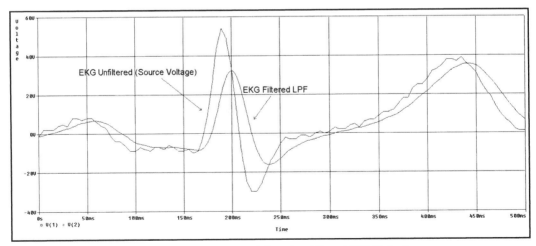

Figure 11.56 — *AMS Simulator* window plotting pure EKG (Input) and filtered EKG (Output).

+ symbol in the first column to indicate a continuation of the previous line.

An interesting circuit to study with Transient Analysis is a half-wave rectifier with a capacitor filter. The inrush current at the time of turn-on is not easily measured in the laboratory without some kind of storage device (analog or digital storage oscilloscope). With *PSpice* Transient Analysis this phenomenon can be easily examined in great detail, and insight can be gained into power supply operation. **Figure 11.57** is the schematic diagram of the rectifier circuit. A Transient Analysis will be carried out for this circuit that will run from 0 ms to 40 ms with a step size of 40 μs.

Figure 11.58 plots the load voltage on the top and the diode current on

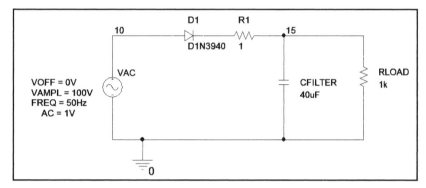

Figure 11.57 — Rectifier circuit schematic.

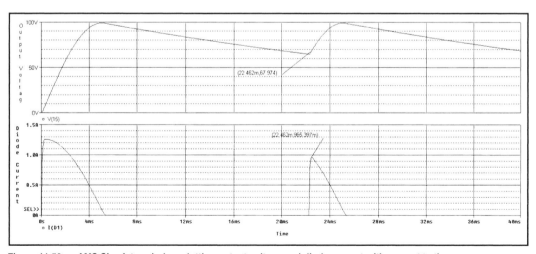

Figure 11.58 — *AMS Simulator* window plotting output voltage and diode current with respect to time.

the bottom. The load voltage rises from 0 V, peaks at nearly 100 V, and has the quasi-sawtooth shape characteristic of a poorly filtered power supply. The ripple voltage is about 33 V_{pp}. Using the cursor, we can see that the diode current during the first positive half cycle (enrich current) peaks at about 1.26 A, while during the second positive half cycle the maximum is 1 A. This is because the inrush current must charge the capacitor, which is initially at 0 V, and raise its voltage to 100 V. On subsequent positive half cycles, the capacitor voltage is never less than 62 V, so less current is needed to raise its voltage back to 100 V.

A complementary metal-oxide semiconductor (CMOS) FET inverter NAND circuit is connected to two square wave voltage sources in the next example, and the output is determined as a function of time. The two square waves will provide all possible logic input conditions (11, 01, 10, 00 binary) to the NAND gate. The logic NAND gate is made with four enhancement MOSFETs; two are N-channel and two are P-channel.

Figure 11.59 shows the circuit, in which the NAND gate output is

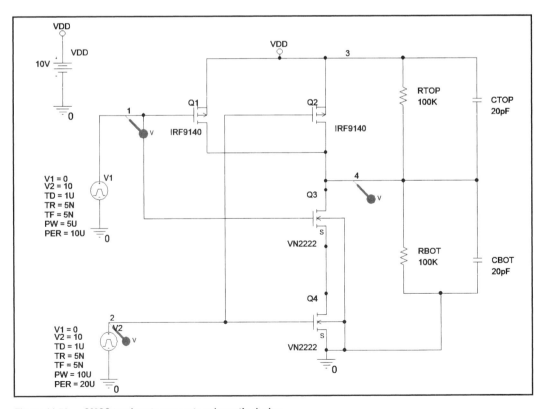

Figure 11.59 — CMOS two input NAND gate schematic design.

loaded resistively and capacitively. The circuit consists of four enhancement-type MOSFETs and two VPULSE sources that output square waveforms that make up a binary input to the NAND circuit. For enhancement-mode MOSFETs, VTO is positive for N-channel devices and negative for P-channel devices. The starting time of the pulses is delayed by 1 μs, and the rise and fall times (5 ns) are quite small compared to the pulse widths (5 μs and 10 μs). A simulation profile is set up to run from 0 ms to 24 μs with a step size of 24 ns.

The analysis results are shown in **Figure 11.60**. The upper plot contains the two input voltages and the lower plot is the NAND output voltage by itself. It can be seen that the NAND gate output is low only when both inputs are high, and the rise and fall times of the output are significant. If desired, the cursor in PROBE could be used to measure rise time and fall time of the output.

A 7404 TTL inverter logic gate is shown in **Figure 11.61**. It is fed by a TTL-compatible square wave (well, nearly square). The output is connected to a pull-up resistor of 2 kΩ. This design uses a configuration of four NPN BJTs connected to a VPULSE input, V2.

Again, this design will use the NPN transistor from the *PSpice* Breakout library. In order to run the simulation, parameters need to be declared in *Model Editor*. Open *Model Editor* as shown in **Figure 11.62**. The parameters for the four transistors are shown in the MODEL EDITOR window (see **Figure 11.63**). Save these changes and run a Transient Analysis on the circuit with a start time of 0 ns and an end time of 105 ns with a step size of 105 ps.

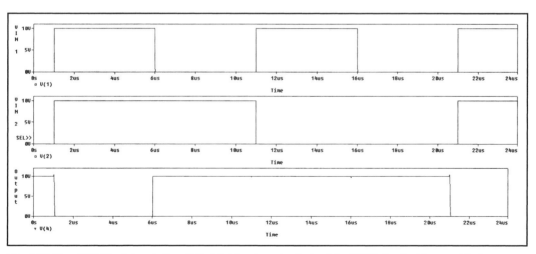

Figure 11.60 — *AMS Simulator* window plotting the inputs and outputs of the NAND circuit with respect to time.

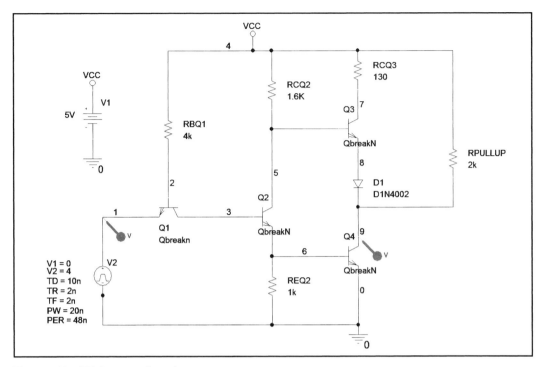

Figure 11.61 — TTL inverter schematic.

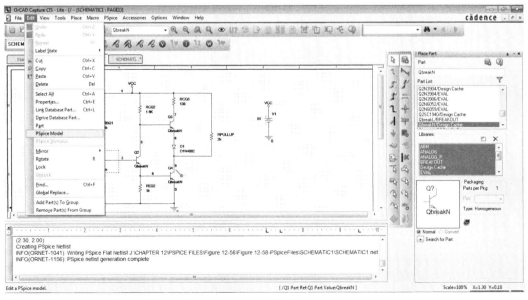

Figure 11.62 — Opening *Model Editor* for the breakout component.

The input file, Figure 11.61, describes the input pulse as going from 0 V to 4 V after a 10 ns delay, with a 2 ns rise time and fall time, a pulse width of 20 ns and a period of 48 ns. The transistor model QBREAKN includes a substantial number of parameters that override the *PSpice* default parameters for a BJT.

The PROBE display is a graph of input and output voltages (see **Figure 11.64**). The output waveform is valid from a logic standpoint, and it shows the distortions that are to be expected from a TTL logic gate operated at 20.8 MHz.

Transient Analyses on astable circuits and free-running oscillators can

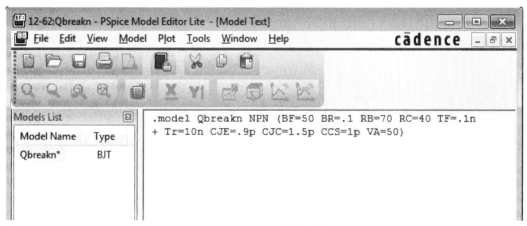

Figure 11.63 — Active MODEL EDITOR window for component **QBREAKN**.

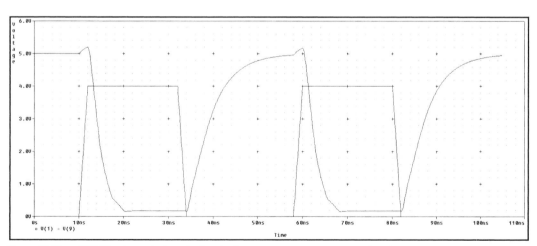

Figure 11.64 — *AMS Simulator* window plotting the input and output waveforms of the inverter circuit.

be a bit tricky to accomplish, and one must pay careful attention to the initial conditions in the circuit. For example, the circuit shown in **Figure 11.65** has complete symmetry (at least on paper), but in reality, at the time power is applied to the physical circuit, one transistor will go to saturation and the other will become cut off. As far as *PSpice* is concerned, it is unknown which transistor will win the race to conduction. In order to help *PSpice* with the transient solution, a good technique is to set some initial conditions to give *PSpice* a starting point for further analysis. This, of course, requires that you have an understanding of how the circuit starts and runs.

Initial conditions include the five node voltages specified in the voltage source parameters, where at time 0 Q1 is on (set by V(1)=.1 and V(3)=0.8) and Q2 is off (set by V(4)=6 and V(5)=–6). Of course, in a physical circuit, slight differences in transistor parameters and/or stray circuit capacitance might just as likely cause the opposite condition. To aid the efficiency of the computation during this simulation, options of the simulation profile will be modified. Parameters of the Transient Analysis will be entered into the SIMULATION SETTINGS window as usual (see **Figure 11.66**).

In this window, select the OPTIONS tab and assign the value of RELTOL to be "0.01" (see **Figure 11.67**). This makes the relative tolerance 1% instead of the default value of 0.1%, which saves some computation time

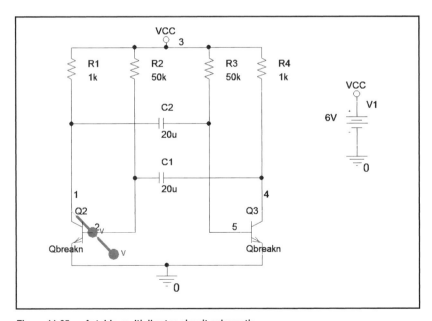

Figure 11.65 — Astable multivibrator circuit schematic.

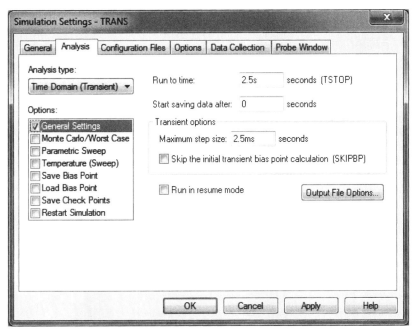

Figure 11.66 — SIMULATION SETTINGS window, Transient Analysis.

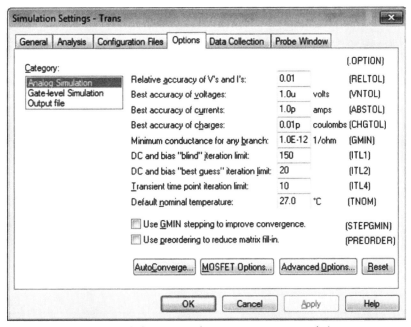

Figure 11.67 — Adjusting relative accuracy in SIMULATION SETTINGS window.

on a circuit such as this with very fast voltage transitions.

The graph in **Figure 11.68** shows the collector voltage of Q1 in the lower plot and the base voltage of Q2 in the upper plot, versus time. The collector voltage is essentially a square wave, and the base voltage shows the exponential charging of the 20 µF timing capacitor connected to it. PROBE's cursor used on the base voltage waveform allows the oscillation period to be measured as 1.4539 s. The behavior of the circuit once free-running oscillation has started confirms the validity of the original initial conditions.

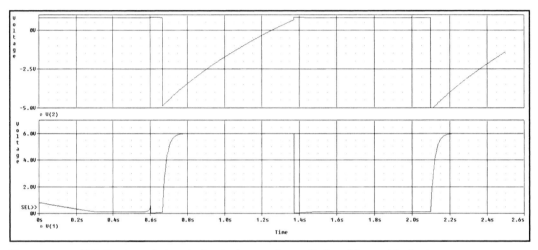

Figure 11.68 — *AMS Simulator* window plotting the free-running oscillation of the astable circuit.

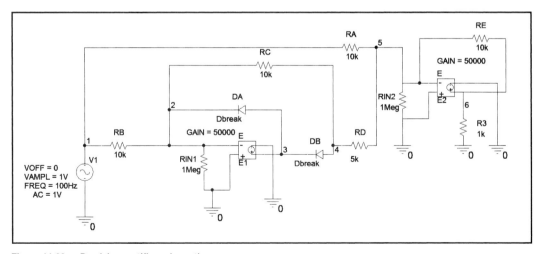

Figure 11.69 — Precision rectifier schematic.

A real diode has an offset or barrier voltage (about 0.6 V for silicon) that must be overcome before forward conduction begins. This makes real diodes unusable by themselves to rectify low-level voltages like audio signals. For rectifying small voltages, another technique must be used to overcome this limitation. **Figure 11.69** is a precision rectifier circuit, sometimes called an absolute value circuit. The input file in Figure 11.69 models the two op-amps using VCVSs with a gain of 50,000 V/V, which is controlled by the differential voltage across nodes 10 and 20 in the subcircuit. Each VCVS essentially models an op-amp (similar to prior examples) with a very high input resistance. The very large open-loop gain of the VCVS-modeled op-amp E11 is used to negate the diode offset voltage.

The diodes are truly ordinary in that they are modeled by the *PSpice* default diode parameters, which gives them an offset voltage typical of silicon. The input voltage is a 1 Vp, 100 Hz sinusoid, and the Transient Analysis occurs for two complete periods of the input. Thus the simulation profile will be set for a total runtime of 20 ms with a step size of 20 µs.

In **Figure 11.70** the input voltage is plotted in the top graph and the output voltage in the one below. The output voltage is indeed the absolute value of the input voltage, with no visible difference between input and output peak voltage caused by a diode. The accuracy of this circuit depends on both the large open-loop voltage gain of the op-amp and the exact matching of the resistors.

Figure 11.71 is the schematic diagram of an amplitude modulation generator, although *PSpice* has a built-in model for a frequency modulation generator (SFFM independent source), it does not have an amplitude

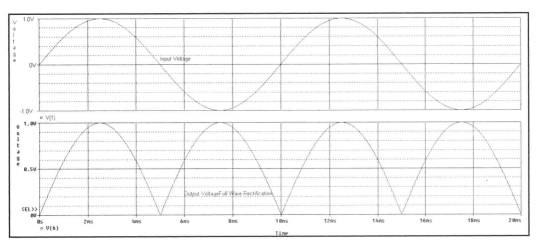

Figure 11.70 — *AMS Simulator* window plotting the full-wave rectification of the input waveform.

modulation generator. By using a MULT component from the ABM (analog behavior model) library, it is possible for a user to take a carrier waveform and multiply it by a modulation waveform. The modulation index can be changed, as can the carrier frequency and modulation frequency. While the SFFM source is limited to a pure sinusoid as the modulation, the AM generator could be modulated by any kind of waveform. A basic Transient Analysis will be carried out on this circuit with a run time of 1 ms and a step size of 1 µs.

The 40 kHz carrier waveform is shown in the top of the output plot in **Figure 11.72**. The 2 kHz modulation waveform is shown as the trace in the

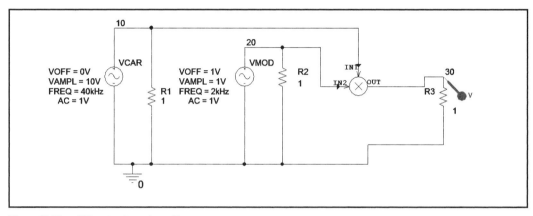

Figure 11.71 — AM generator schematic.

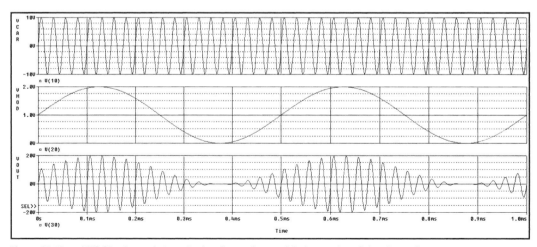

Figure 11.72 — *AMS Simulator* window plotting the carrier, modulation, and modulated waveforms.

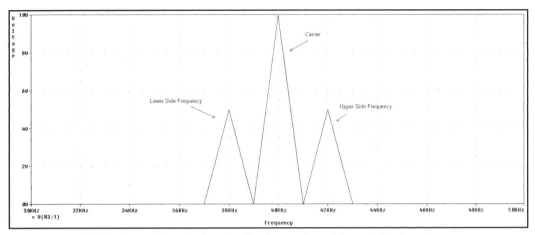

Figure 11.73 — *AMS Simulator* window plotting a fast Fourier transform of the modulated waveform.

middle plot, and the amplitude-modulated waveform is shown in the bottom plot. Remove all of the plots from the *PSpice* window, aside from the AM waveform from the bottom plot. With only the AM waveform displayed, click the FFT button in the top of the *AMS Simulator* window. Clicking the FFT button performs a fast Fourier transform (FFT) and plots the component frequencies of the output waveform. In **Figure 11.73**, the FFT is visible. The carrier amplitude is in the center of the plot, accompanied by the upper and lower side frequencies present as a result of the 2 kHz modulation frequency.

In Summation

This chapter puts together all aspects of the design and simulation procedures covered in the previous chapter in an effort to demonstrate the capabilities of the *OrCAD/PSpice* software through practical applications. The last chapter of this book takes these procedures a step further and begins to explore some of the more advanced features of the Cadence software package.

Chapter 12

Advanced Analysis

Until this final chapter, the text has focused mainly on carrying out four basic simulation types on various circuit designs and interpreting the simulation results from the Cadence *Allegro AMS Simulator* or *PSpice A/D* software. From a design standpoint, these simulation types are essential to lay the proper groundwork for a given circuit, but it is important to remember that any simulation results yielded by these designs are based on the behavior of an ideal *PSpice* model. The Transient Analysis, DC Sweep, AC Sweep and Bias Point Calculation can characterize the general behavior and functionality of a circuit, but they do not take into consideration component tolerances, environmental factors (such as temperature), and noise generated from real components.

The *PSpice* software has the capability to take all of these factors into consideration and give you an idea of what you might expect from a particular component choice and all of the possible ways it can impact the performance of a circuit. These types of simulations fall into the realm of what we will call Advanced Analysis (AA).

There are five types of AA simulations that will be discussed in this chapter. They are Sensitivity Analysis, Monte Carlo Analysis, Smoke Analysis, Noise Analysis and Temperature Analysis. The point must be made that all of these simulation types are so immensely complex that each one could have multiple chapters devoted to it. The purpose of this chapter is to define their most primary functions and give the user some insight by using the most basic examples.

Advanced Analysis Schematic Preparation

To learn how to carry out these types of simulation procedures, we will refer to the BJT amplifier that we biased at the end of Chapter 7 with the help of the Cadence software. But before we can explore these simulation techniques, we will need to make some modifications to the design. The circuit has been redrawn in a new project and is shown in **Figure 12.1**.

In order to carry out some of the AA simulation procedures, we will need to incorporate tolerance parameters into the design. This is quite simple with a majority of the passive components in the *PSpice* library, such as resistors, capacitors and inductors. Active components with more complex *PSpice* models require the placement of AA components. In the case of the circuit shown in Figure 12.1, all the resistors and capacitors remain in place. The 2N3904 will be removed and substituted with a 2N3904 equivalent model from the AA *PSpice* library. To do this, click the PLACE PART button in the *OrCAD Capture* or *Allegro Design Entry* workspace to expand the PLACE PART menu to the right of the screen and click the button to add a new Component library, just as if you were adding a new Component library to a project. In the *PSpice* library folder you should see all of the familiar libraries that you have been using thus far, but this folder also contains a folder with the title ADVANLS (for Advanced Analysis). This folder contains all of the AA libraries (see **Figure 12.2**). Open this folder and you will see all the AA libraries that are available to you. Add these libraries to your project to utilize these components in the *OrCAD* workspace.

After adding these libraries to the open project, the AA 2N3904 can be added into the existing schematic design. **Figure 12.3** shows the AA 2N3904 selected in the part list. At this point, you may be wondering how to designate between a standard 2N3904 model and the AA 2N3904. With

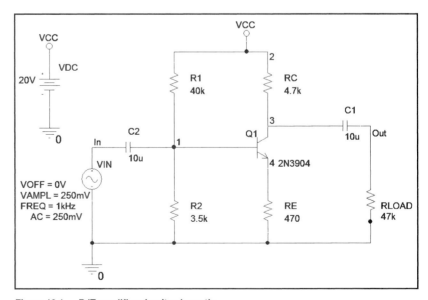

Figure 12.1 — BJT amplifier circuit schematic.

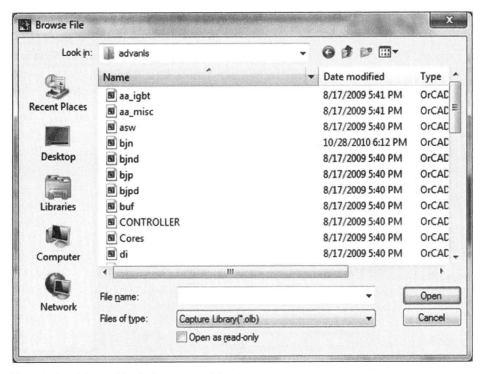

Figure 12.2 — Advanced Analysis component libraries.

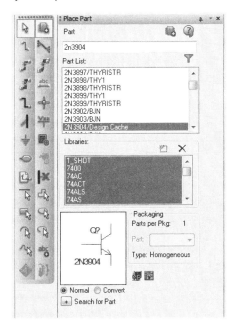

Figure 12.3 — Advanced Analysis 2N3904 selected in PLACE PART window.

the part selected in the part list, look for the *AMS Simulator* icon at the bottom of the screen just to the right of the schematic symbol. As you recall from Chapter 3, the presence of the *AMS Simulator* icon indicates that the schematic symbol to be used in the *Allegro Design Entry* or *OrCAD Capture* circuit has a *PSpice* model. In the case of an AA component, the *AMS Simulator* icon is overlaid with an uppercase P. This is seen with the QP designation for the 2N3904 shown in Figure 12.3.

In **Figure 12.4** this component has been substituted in place of the standard 2N3904 from the original design. The schematic symbol is virtually identical to the original component. The difference lies in the model data that will be used during the simulations.

At this point, the schematic design is set, and now tolerance parameters need to be attached to the components within the circuit. AA simulations will consider component tolerances only if you declare them. If a user neglects to declare a tolerance for a component, the software will use the ideal model for any simulation data. For the purposes of the examples to be completed in this chapter, tolerances will be assigned to the biasing resistors R1 and R2, the collector resistor RC, the emitter resistor RE, and the beta factor of the transistor. With these new parameters in place we will determine if any variation in these tolerances (of R1, R2, RC, or RE) will affect the integrity of the bias of this BJT circuit.

To begin, let's assume that we would like this circuit to be constructed using resistors with a 10% tolerance. To assign a tolerance to these resistors,

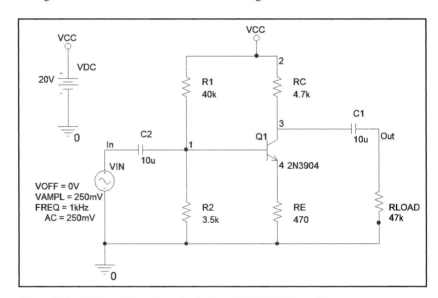

Figure 12.4 — BJT amplifier schematic design with AA 2N3904 model.

select R1, R2, RC, and RE in the *OrCAD Capture* schematic capture window and double-click on them to open the Property Editor. There is a column within the Property Editor labeled TOLERANCE. Under this column, declare the desired tolerance percentage for each component. In **Figure 12.5**, the 10% tolerance has been declared for each of the four resistors.

After applying the changes, with the TOLERANCE column still selected, you can click the DISPLAY button to label this parameter in the *OrCAD* workspace. This has been done in **Figure 12.6**.

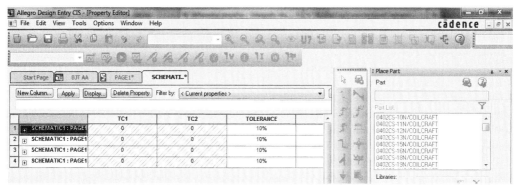

Figure 12.5 — PROPERTY EDITOR window displaying tolerances.

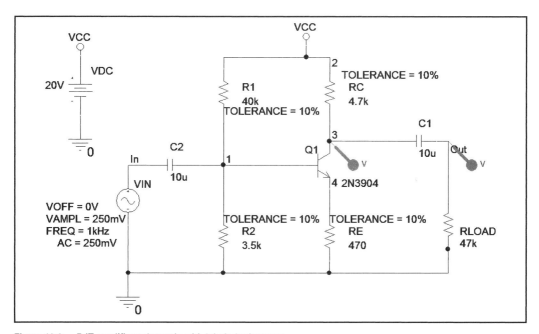

Figure 12.6 — BJT amplifier schematic with labeled tolerances.

We have now declared tolerances for each resistor. The last step will be to declare a tolerance for the beta factor of the transistor. To do this, right-click on the transistor and select EDIT PSPICE MODEL to open *Model Editor* (see **Figure 12.7**).

When using *Model Editor* with the standard *PSpice* components in previous chapters, you might recall that *Model Editor* would dictate component characteristics in a text format. All of the alterations made to the examples in Chapters 10 and 11 strongly resembled a text editor window. With the AA components, the model data opens into an interactive spreadsheet. This is shown in **Figure 12.8**.

Notice in the MODEL EDITOR window the presence of two columns, POSTOL and NEGTOL. These columns indicate the percentage tolerance that the beta factor can sway in the positive and negative direction, respectively, from the defined beta value. In Figure 12.8 it has been declared that the

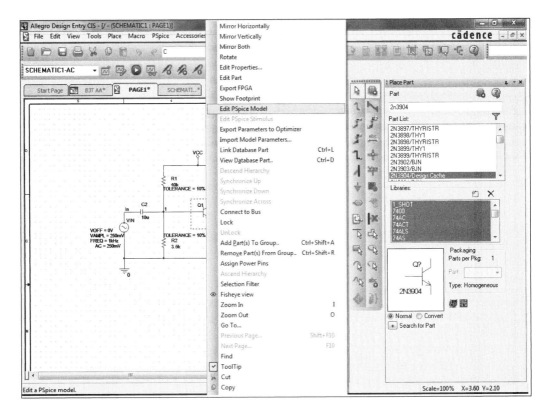

Figure 12.7 — Opening *Model Editor* for an AA component.

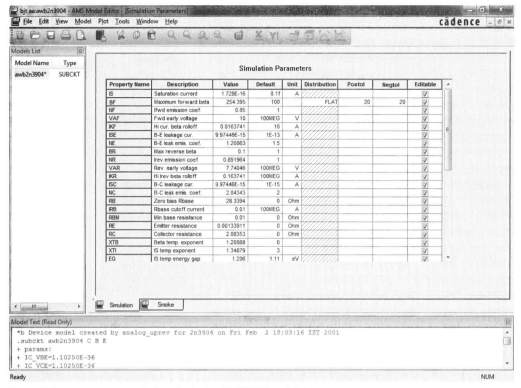

Figure 12.8 — MODEL EDITOR window displaying AA 2N3904 model data.

tolerance of the transistor can deviate 20% in the positive direction and 20% in the negative direction. The 2N3904 transistor model has a beta value defined at 254.395. With these tolerances in place, an AA can predict variations in the behavior of this amplifier circuit with arbitrary beta values that fall within the range of 203.516 to 305.274. It can be noted that these POSTOL and NEGTOL columns are modifiable for all component parameters. In this example, a tolerance has been assigned only to the beta factor. In actuality, the results of an AA simulation would be more closely representative of a real component if one were to assign tolerances to more of these model parameters.

Review of Ideal Circuit Behavior

Our schematic has been fully prepared and can now undergo the AA simulations. Before we do so, it is important to determine the general behavior of the circuit with the declared values. This will establish a baseline to help interpret any data that will be obtained from future AA simulations. A Bias Point Calculation, Transient Analysis, and AC Sweep will be conducted. These basic simulations will not take the declared component tolerances into consideration. **Figure 12.9** shows a Bias Point Calculation with all of the node voltages labeled.

Next we will set up a simulation profile for a Transient Analysis. The sinusoidal voltage source VIN is operating at 1 kHz. A run time has been set at 1 ms to display one cycle of the output waveform, and a step size has been set at 1 μs. These simulation settings are shown in **Figure 12.10**.

Figure 12.11 shows a BJT class A amplifier. With a 250 mV input signal applied to the input, an output voltage of approximately 2.167 V is measured across the load resistor. It is important to remember that this voltage is measured across the load resistor. We have not associated a tolerance with this load resistor, but for all the sample simulations in this chapter we will assume that the load is ideal and fixed at 47 kΩ. To compensate for this, an additional marker has been placed on the other side of the coupling capacitor on node 3, the output of this amplifier at the collector terminal, and the trace

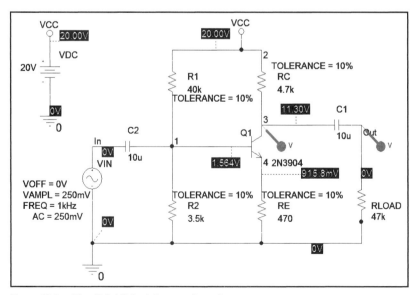

Figure 12.9 — Bias Point Calculation results, voltage.

has been added to the plot shown in **Figure 12.12**.

Obviously, the same gain can be observed in both waveforms. The key difference is the dc offset seen with the trace of the node 3. Referring to the Bias Point Calculation, this voltage is determined to be 10.93 V. During the AA simulations, using node 3 as a reference point, we will assess the

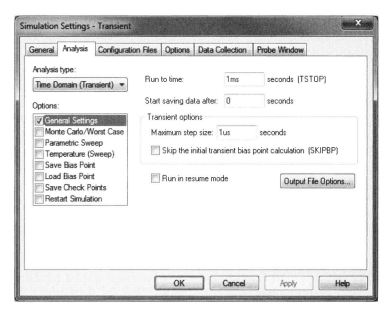

Figure 12.10 — SIMULATION SETTINGS window, Transient Analysis.

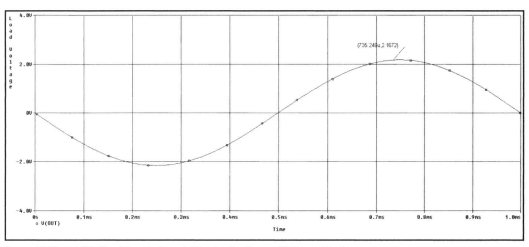

Figure 12.11 — *AMS Simulator* window plotting load voltage with respect to time.

stability of the bias of the BJT amplifier as component values fluctuate due to tolerances.

Lastly, an AC Analysis is performed on the circuit to determine the frequency response of this BJT amplifier. The simulation parameters can be seen **Figure 12.13**. The sweep begins at 10 Hz and ends at 100 MHz with 100 data points per decade plotted along a logarithmic axis.

The load voltage is measured during this frequency sweep and is shown in the plot of **Figure 12.14**. A data point has been marked to again verify voltage gain.

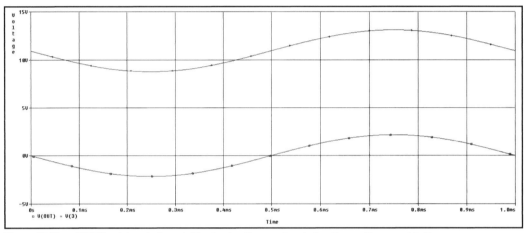

Figure 12.12 — *AMS Simulator* window plotting load voltage and node 3 voltage with respect to time.

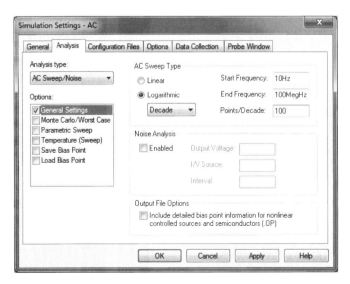

Figure 12.13 —
SIMULATION SETTINGS
window, AC Analysis.

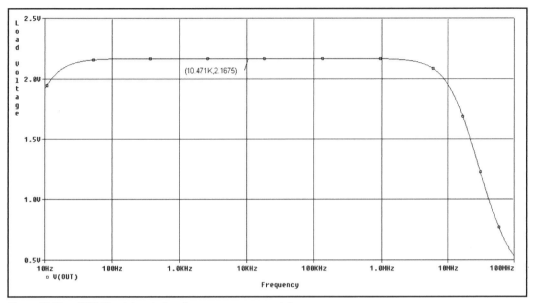

Figure 12.14 — *AMS Simulator* window, frequency response plot.

Sensitivity Analysis

The first simulation type to be discussed is the Sensitivity Analysis. The Sensitivity Analysis will carry out a series of runs on a circuit while varying the values of components within their declared tolerance ranges. (Note that in the *Lite* — demo — version of the software there are limitations on the number of runs that can be used. You may need to set the number of runs to a value lower than that used in the examples in this chapter.) Let's begin by opening the AA window. To do this, select ADVANCED ANALYSIS, then SENSITIVITY under the *PSpice* drop-down menu. This is shown in **Figure 12.15**.

This opens the AMS ADVANCED ANALYSIS window (see **Figure 12.16**). Most of the Sensitivity Analysis simulations discussed in this chapter utilize this window. You will immediately notice that the drop-down menu at the top of the screen reads SENSITIVITY. In the *OrCAD* window, had the user selected Monte Carlo, or Smoke, this same window would have opened; however, the user interface would look slightly different based on the selected simulation type. Also, it is important to note that all of the parameterized components are listed in the upper portion of the window. The lower portion of the window is devoted to measurements and project specifications.

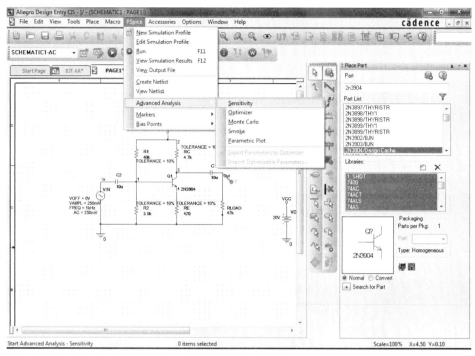

Figure 12.15 — Opening *PSpice* Advanced Analysis, Sensitivity.

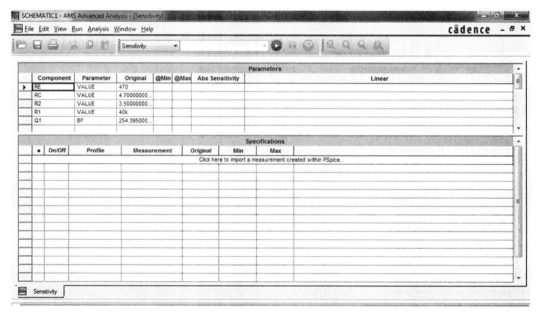

Figure 12.16 — AMS ADVANCED ANALYSIS window, Sensitivity.

First, we will demonstrate how to declare a measurement; then we will discuss how this measurement is relevant to the Sensitivity Analysis, what the Sensitivity Analysis is, and its applications. Earlier, it was stated that we will be using the voltage at node 3 as a reference and would monitor the stability of the bias as component values varied. Therefore, for our Sensitivity Analysis, we use the same measurement V(3), the voltage at node 3, that was seen in the Bias Point Calculation and used as a trace measurement for the Transient Analysis and AC Sweep.

To add this V(3) measurement to the Sensitivity Analysis window, select CREATE NEW MEASUREMENT under SENSITIVITY under the ANALYSIS drop-down menu. This is shown in **Figure 12.17**.

The EDIT MEASUREMENT window (see **Figure 12.18**) bears a remarkable resemblance to the ADD TRACE window in the *AMS Simulator* software. In the top left corner of the window, there is a PROFILE drop-down menu. Earlier in the chapter, three simulations were carried out on the BJT amplifier — a Transient Analysis (Simulation Profile Title, transient.sim), AC Sweep (Simulation Profile Title, ac.sim), and a Bias Point Calculation.

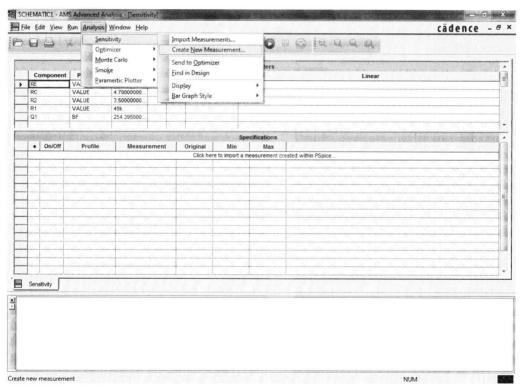

Figure 12.17 — Creating a new reference measurement.

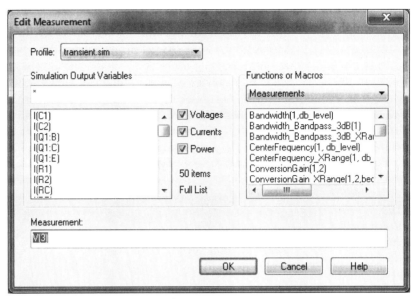

Figure 12.18 — EDIT MEASUREMENT window.

These profiles exist in this drop-down menu. Transient.sim has been selected in Figure 12.18. With the transient simulation profile selected, all of the available measurements (V, I, and W) that existed in the ADD TRACE window are now visible in the column browser of the EDIT MEASUREMENT window. The measurement V(3) can either be found and selected in the column browser, or it can be entered manually into the measurement text box. Click OK to add this measurement. The measurement text box can also be used in conjunction with the functions and analog operators at the right of the EDIT MEASUREMENT window to perform any measurement imaginable to evaluate a circuit's performance. You also may have noticed the IMPORT MEASUREMENT feature under the analysis menu shown in Figure 12.17. This method allows you to directly import a trace function saved from an earlier simulation. This can be especially handy if you are dealing with a complicated function from the *AMS Simulator* and simply want to carry that function over into the AMS ADVANCED ANALYSIS window.

Now that our measurement is in place, we must define the purpose of the Sensitivity Analysis. The Sensitivity Analysis will determine which components of a design have values that are the most critical to the performance of the circuit. In other words, if each component value is varied within its tolerance range, which will yield the most significant change in the reference measurement declared in the EDIT MEASUREMENT window.

Remember, the BJT amplifier has four resistors with tolerances and the transistor itself has a tolerance for its beta factor. The Sensitivity Analysis will determine this and can present the data in a variety of ways.

The two types of sensitivity results that will be presented in this chapter are Absolute Sensitivity and Relative Sensitivity. The easiest way to define each interpretation of the data is to also offer a series of examples, as follows:

• Absolute Sensitivity — A ratio that illustrates the amount of change in a measured value due to an incremental change in a component value

Example — If there is a 1 Ω change in resistance, there will be a 0.7 V change in output voltage.

• Relative Sensitivity — The percent change seen in a measured value if a component value is varied by 1%.

Example — If there is a 1% change in resistance, there will be a 4% change in output power.

With the measurement in place, to run the Sensitivity Analysis, click the RUN button at the center of the toolbar at the top of the screen. The results of the Absolute Sensitivity Analysis are shown in **Figure 12.19**.

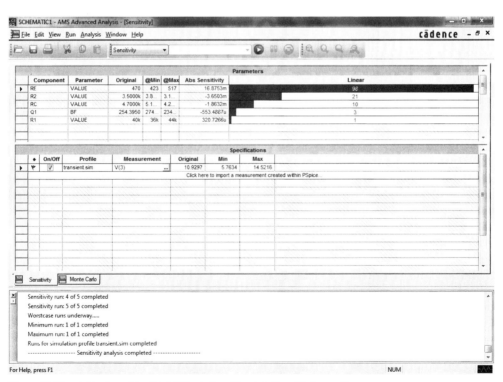

Figure 12.19 — AMS ADVANCED ANALYSIS window, Absolute Sensitivity results.

Referring to our definition of the Absolute Sensitivity, we know that these results are a function of the simulation software adjusting resistances and the beta factor in incremental values and determining the change in the measurement that was set as V(3), the voltage at node 3. After running the simulation, the maximum and minimum calculated component values will appear in the columns in the upper half of the window along with a bar graph representing the ratio of the change in voltage with respect to the change in a component's parameter.

Based on these results, it can be concluded that an incremental change in the resistance of RE, the emitter resistance, will yield the most significant change in the voltage at node 3. This ratio stands out far beyond the rest of the components being evaluated, the next being R2 with a ratio of 21, in comparison to 98. At the other end of the spectrum, it can be noted that an incremental change in R1, or the beta value of the transistor, will yield a minimal change in voltage at node 3.

After clicking RUN, the *PSpice* or *Allegro AMS Advanced Analysis Simulator* will perform sensitivity runs on each of the parameterized components in the circuit. These runs produce the bar graph seen at the top of the screen. These sensitivity runs are followed by what are called worst-case runs. Worst-case runs determine the greatest changes in the user-defined measurements (in this case, V(3)) and calculate maximum and minimum values. The bottom half of the screen displays the measurement V(3). To the right of this measurement there are three data columns. The first column is the original value (10.9297V) of voltage at node 3, calculated using the exact component values. This is the same value of voltage that was seen earlier in the chapter with the Bias Point Calculation and in the dc offset of the Transient Analysis. The next two columns, MIN and MAX, show the largest possible deviation from the measured value that could possibly occur with components values being at the limits of their tolerance. In this example, component tolerances could potentially yield a minimum voltage of 5.7634 V and a maximum voltage of 14.5216 V at node 3.

Right-click on the bar graph and select RELATIVE SENSITIVITY. The bar graph and columns will change to depict the results of the Relative Sensitivity Analysis. These results can be seen in **Figure 12.20**. Remember, the key difference between Absolute and Relative Sensitivity Analysis is that Relative is based on a percentage change in a component's parameter, as opposed to an incremental value. When each component varies a single percentage within its tolerance range, R2 and R1 seem to have the greatest impact toward the change in V(3). This could be because a percentage change in a large resistor like R1 accounts for a significant change in resistance. When varying the values of R1 and R2, the voltage divider that is biasing

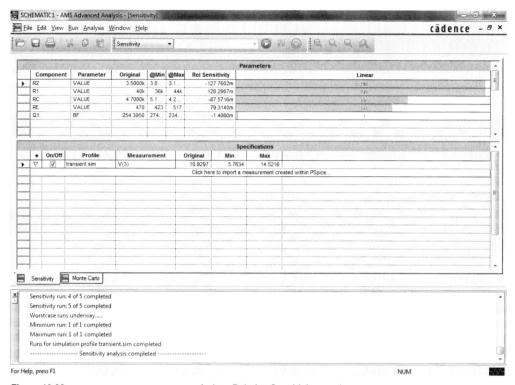

Figure 12.20 — AMS ADVANCED ANALYSIS window, Relative Sensitivity results.

this common emitter configuration is directly affected as well. That would have a significant impact on the behavior of the circuit and could certainly have an impact on the voltage measurement at node 3.

When comparing the results of the Relative Sensitivity Analysis to the Absolute Sensitivity Analysis, it makes perfect sense that the smaller resistor values, RE and R2, would have the highest ratios of change in V(3). This is because a 1 Ω variation in the resistance of a smaller resistor like RE or R2 accounts for a larger percentage of change in resistance in comparison to a 1 Ω shift for a larger resistor like R1. One last note on the Sensitivity Analysis is that results can also be presented in a logarithmic scale. With either the Absolute or Relative results displayed, simply right-click on the graph and select the logarithmic option.

Monte Carlo Analysis

Next we will move on to the Monte Carlo Analysis. The Monte Carlo Analysis predicts the statistical probability that a certain measurement will be achieved throughout a set number of runs (simulations) where each run will have random component values within their respective tolerance ranges. To begin, expand the drop-down menu at the top of the screen that currently reads SENSITIVITY and change it to MONTE CARLO. The AMS ADVANCED ANALYSIS window will change and look similar to what is shown in **Figure 12.21**.

Before we go any further, we need to create a measurement for reference. We will use the same reference point and type of measurement as was used for the Sensitivity Analysis in the previous example — the voltage at node 3. Add this measurement with the same procedure used for the Sensitivity Analysis. Open the EDIT MEASUREMENT window by selecting CREATE NEW MEASUREMENT under MONTE CARLO in the ANALYSIS drop-down menu. The very same EDIT MEASUREMENT window will appear, except this time you will be adding the V(3) measurement to the MONTE CARLO window. This is shown in **Figure 12.22.**

Next we will modify some of the parameters. Under the EDIT menu select the PROFILE SETTINGS option. When the window opens, select the MONTE CARLO tab to display a window similar to the one shown in **Figure 12.23**.

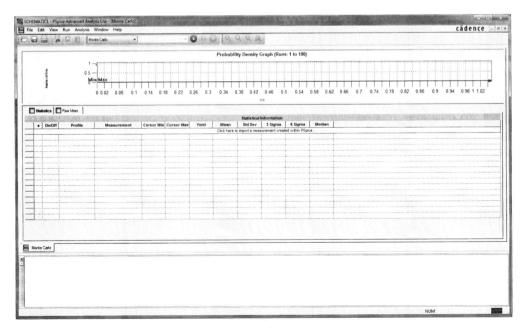

Figure 12.21 — AMS ADVANCED ANALYSIS window, Monte Carlo.

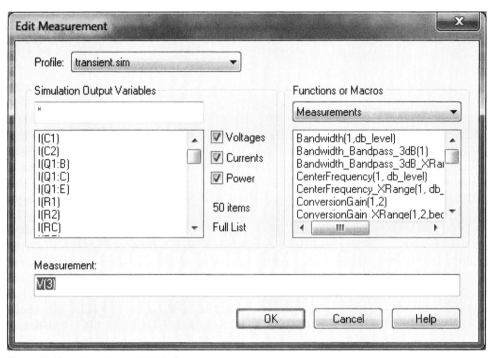

Figure 12.22 — EDIT MEASUREMENT window.

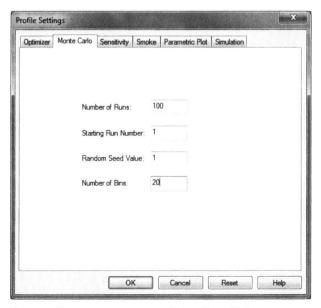

Figure 12.23 — PROFILE SETTINGS window, Monte Carlo.

We will increase NUMBER OF RUNS to 100. This means that the circuit will be simulated 100 times, each time with different component values. (Note that the "Lite" version of the software is limited to fewer runs.) Also increase NUMBER OF BINS to 20. The Monte Carlo Analysis will present the results of these calculations using a histogram. The bin number means that the range of V(3), plotted along the X axis, will be subdivided into 20 equal regions. Each bin will be represented by a single bar in the histogram. The Y axis can plot either the number of runs or the percentage of runs, depending on how the user decides to configure the plot. The default setting will plot the number of runs on the Y axis. This can be changed by simply right-clicking the Y axis and selecting the alternative.

Click the RUN button to begin the simulation. The number of runs and the complexity of the schematic design can have a significant impact on the time that it takes to complete the simulation, so it may take a few moments. What is happening during each run is a random selection of component values. The simulation software performs a Small Signal Bias Solution and extracts only the relevant data. In this case, the relevant data is the measurement that was just created — the calculated voltage at node 3. That calculated data falls into one region/bin of the X axis, and after all of the runs are complete, the collective run data is calculated and plotted in each region of the histogram. This type of graph is referred to as a Probability Density Function graph, often abbreviated as PDF.

The results of the simulation are shown below in **Figure 12.24**. A total of 100 runs were completed. According to the PDF graph, 12 of the runs with randomized components produced circuits that would achieve a voltage ranging from 9.82 V to 10.10 V. The greatest probability falls within this region. With random component values, the circuits tested produced a range of V(3) at a minimum of 8.1 V and a maximum of 13.74 V. This type of simulation can be an especially important tool that may be used during the manufacturing process. When fabricating a real circuit and using components with the tolerances defined in this design, it would be beneficial to have a prediction of how many circuits would meet a certain specification. Depending on the results of the PDF plot, an engineer might want to further refine the tolerance of certain components in a design to produce more circuits within a set of random runs that would meet a particular specification.

The Monte Carlo Analysis results can be viewed using a different interpretation of the data. The plot shown in Figure 12.24 is a Probability Distribution Graph and is the default output of the Monte Carlo Analysis. Right-click on the graph and select CDF GRAPH. CDF stands for Cumulative Distribution Function, and from a mathematical standpoint, this graph is nothing more than the integral of the PDF.

This graph is shown in **Figure 12.25**. To further expand upon this

Figure 12.24 — AMS ADVANCED ANALYSIS window, Monte Carlo PDF.

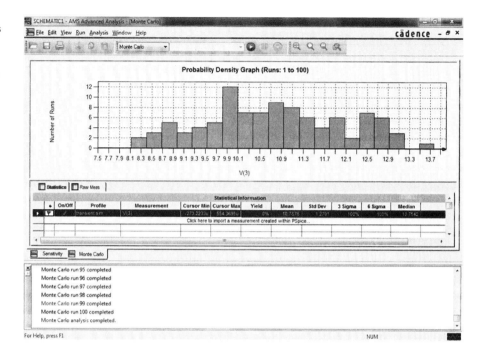

Figure 12.25 — AMS ADVANCED ANALYSIS window, Monte Carlo CDF.

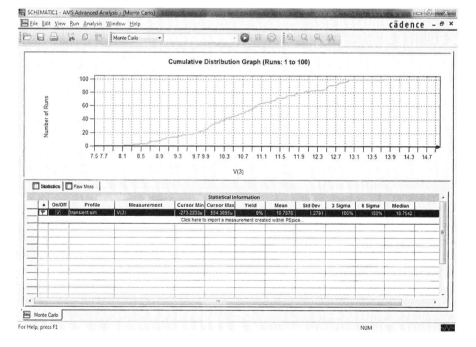

definition, we can examine a few data points and determine exactly what they mean. The X axis of this graph remains V(3), and the Y axis is still set as the number of runs. Let's pick a point on the X axis that was referenced while talking about the PDF plot — 10.1 V. Looking at the CDF graph, and using the cursor, this point corresponds with 34 runs on the Y axis. This data is interpreted as 34 out of 100 runs of yielded voltage measurements at node 3 that were 10.1 V or less. Now we can refer to Figure 12.24, the PDF graph. Starting at the bin with the upper boundary of 10.1 V and descending, the total sum of the quantities of each bin is equal to 34. Looking to the left and right boundaries of the CDF plot, the maximum and minimum values correspond with the outer regions of the maximum and minimum bins of the PDF plot.

Temperature Analysis

The next type of simulation, the Temperature Sweep, technically is not considered an AA simulation method in the most recent version of Cadence, but it is slightly more complex than the standard simulation procedures.

We know that electrical components will exhibit different behavior under different temperature conditions, and *PSpice* will simulate all circuits at a default value of 27 °C. This default setting can be changed under the OPTIONS tab in the SIMULATION SETTINGS window (see **Figure 12.26**). This poses a question as to whether it is possible to observe how a measurement will change as the environmental temperature varies. This can be studied

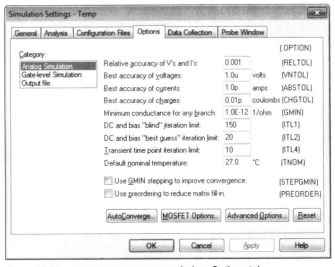

Figure 12.26 — SIMULATION SETTINGS window, Options tab.

through a variation of the DC Sweep simulation type. As you may recall from Chapter 3 and our discussion of a particular component's element line, there exists a temperature coefficient that dictates change in a component's parameter as the temperature of the environment changes.

Before going any further, we must make sense of this temperature coefficient. The temperature coefficient (TC) reads from the element line as follows:

TC = TC1, TC2

Example: Default Value, TC = 0, 0

Recall that the default temperature of any simulation is at the nominal default of 27 °C. The mathematical formula that interprets this temperature coefficient is given as:

$$R_{EnvTemp} = R_{NomTemp} [1 + TC1 (EnvTemp - NomTemp) + TC2 (EnvTemp - NomTemp)^2]$$

where

NomTemp = Nominal Temperature

EnvTemp = Environmental Temperature

For example, assume there is a resistor with a value of 350 Ω in the nominal temperature of 27 °C with a temperature coefficient of TC = –0.004, 0. Using the formula given above, if the environmental temperature were 127 °C, the resistance would be 210 Ω.

If the temperature coefficient is not modified, the resistance will remain fixed throughout a Temperature Sweep simulation. It can also be noted that

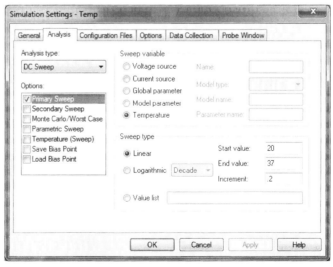

Figure 12.27 — SIMULATION SETTINGS window, Temperature Sweep.

standard transistors and AA transistors may have temperature characteristics associated with their default models. We will illustrate this in the following example.

To sweep the temperature of a circuit, return to the *OrCAD* window and create a new DC Sweep simulation profile. As shown in **Figure 12.27**, select TEMPERATURE as the sweep variable. At the bottom of the window, declare the sweep parameters as you would with a typical DC Sweep. For this example, a sweep will take place with a starting temperature of 20 °C that ends at 37 °C with incremental data points every 0.2 degrees. Apply these simulation parameters and run the simulation.

Like the AA simulations done in earlier examples, add a trace for the measurement of the voltage at node 3. This will plot a change in the node voltage with respect to temperature as a result of the temperature parameters of the AA 2N3904 transistor model. **Figure 12.28** shows linear function of the transistor's temperature parameter. At 20 °C, the voltage at node 3 is approximately 10.98 V. Continuing across the X axis, at the temperature 27 °C we see the voltage at the nominal temperature bias is 10.93 V. This is the same voltage that was seen in the Transient Analysis simulation and the Bias Point Calculation. Then at the far end of the sweep, at 37 °C, the voltage continues its linear decrease to slightly less than 10.86 V. This trend will continue. If the temperature continues to fall, the voltage at node 3 will also continue to fall.

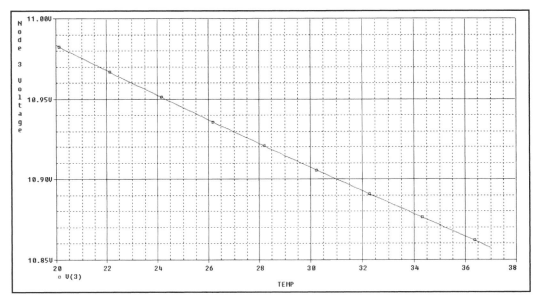

Figure 12.28 — *AMS Simulator* window, Temperature Sweep.

Noise Analysis

The next type of simulation also is not considered an AA simulation method in the most recent version of Cadence, but is extremely important and will be able to help us assess the effectiveness of a design and ensure that this design will produce a quality output. This simulation method is Noise Analysis, and it expands upon a basic AC Sweep simulation. Return to the *OrCAD* window and create a new simulation profile for an AC Sweep. In the SIMULATION SETTINGS window, set the sweep to start at 10 Hz and continue to 100 MHz with a logarithmic scale and 100 data points per decade (see **Figure 12.29**). In the bottom half of this window, check the box to enable the NOISE ANALYSIS. The first text box in this region declares the voltage output of the circuit. The net alias for the node above the load resistor in the amplifier has been labeled "Out", so in this text box, simply enter "V(Out)."

The next text box declares the reference name for the voltage or current source supplying the input signal. This feature makes this simulation type incredibly versatile when considering the fact that certain circuits require more than one analog voltage or current source. Take a differential amplifier for example. If this differential amplifier has two inputs, two separate simulations can be carried out to determine the amount of noise contributed at each input. The reference name for the VSIN source for this circuit is "VIN".

Lastly, the user must enter an interval. The interval parameter can be thought of as similar to the way the user defined the resolution of a Transient

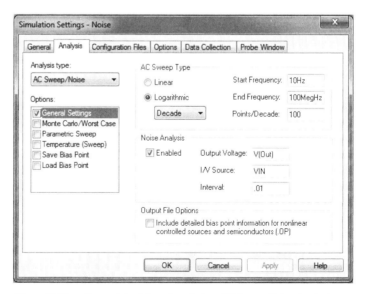

Figure 12.29 — SIMULATION SETTINGS window, AC Sweep/Noise Analysis.

Analysis with the step size. An interval of 0.1 would produce a plot with a very high resolution, as a noise data point would be plotted at every 0.1 Hz. All of these parameters have been entered into the simulation profile and are shown in Figure 12.29.

Run the simulation and once the *AMS Simulator* window appears, enter the traces:

"- V(INOISE)" — Trace name for input noise voltage

"- V(ONOISE)" — Trace name for output noise voltage

These traces will be plotted along the X axis within the range declared in the AC Sweep settings. This output is shown in **Figure 12.30**.

It is easier to interpret these results in decibels (dB), as noise is typically expressed in terms of dB when discussing a circuit. This can be achieved using the analog operators and functions available in the ADD TRACE window. Double-click on each of them to modify the trace names. Adding the DB() trace function to each measurement will perform a dB calculation on both the input noise voltage and output noise voltage. This is shown in the ADD TRACES window shown in **Figure 12.31**. This changes the Y axis of the plot to a dB scale, as shown in **Figure 12.32**.

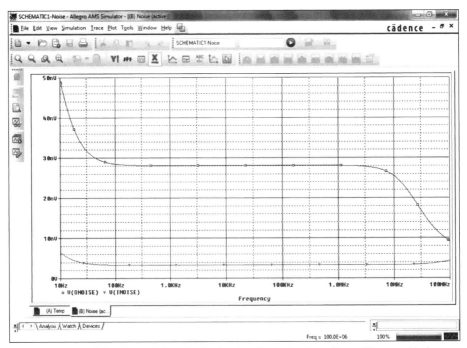

Figure 12.30 — *AMS Simulator* window plotting input and output noise voltage with respect to frequency.

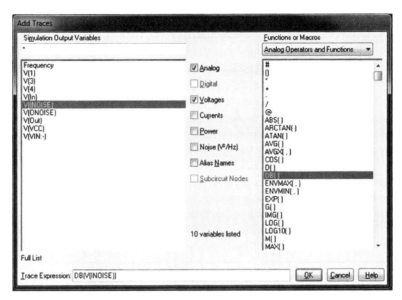

Figure 12.31 — ADD TRACES window, dB trace function.

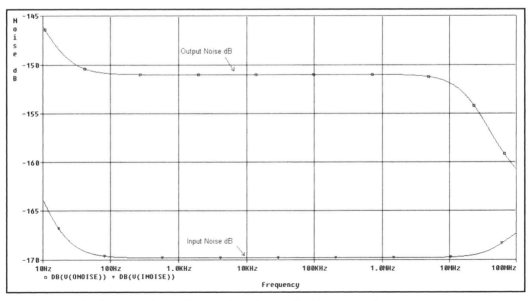

Figure 12.32 — *AMS Simulator* window plotting noise voltage in dB with respect to frequency.

It would also be easier to put these values into perspective if these traces were plotted alongside the results of the AC Sweep. When the noise simulation runs, an AC Sweep is also carried out. To view these results, simply add a second plot to the window and add a trace plotting V(Out). This is shown in **Figure 12.33**. Looking at the X axis, starting at 10 Hz and moving right, as the output across RLOAD rises to its maximum value, the output noise drops to approximately –152 dB and the input to –170 dB. As the frequency increases, approaching approximately 10 MHz, the output voltage begins to roll off from its marked 2.1675 V peak value, as does the output noise. During the roll off of the output voltage a slight increase in the input noise is observed where the input noise nears approximately –167 dB.

Even though this simulation is not considered an AA, it can be an important tool in assessment of a design and can be used in conjunction with AA simulations. Earlier, it was mentioned that it is possible to import measurements from the *AMS Simulator* to the AA simulation windows to be used to evaluate design performance. The traces DB(V(INOISE)) and DB(V(ONOISE)) can be imported into Sensitivity Analysis simulations or Monte Carlo simulations to determine how parameterized components can impact input and output noise, and to predict the probability of noise occurrence over a number of runs.

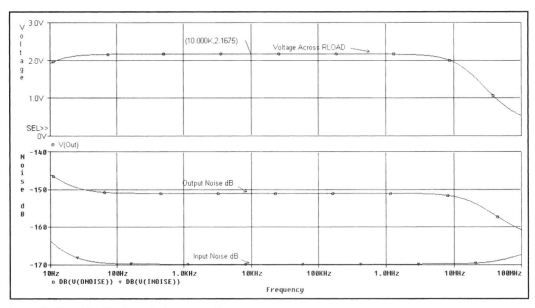

Figure 12.33 — *AMS Simulator* window, noise plot.

Smoke Analysis

For the last simulation type to be discussed in this chapter, return to the ADVANCED ANALYSIS window. In this example we will perform a Smoke Analysis on the BJT amplifier design. The Smoke Analysis will evaluate the stress on all the components in a circuit due to power dissipation, temperature, excessive current, or voltage. To begin, open the drop-down menu in the toolbar at the top of the window and select SMOKE. The window shown should resemble that shown in **Figure 12.34**.

To perform these assessments, the Cadence software utilizes manufacturer operating conditions (MOCs) and derating factors. These two factors are used to determine whether all the components of a circuit will operate under safe operating limits.

Before running the Smoke Analysis, it is important to implement a derating factor. A derating factor is a percentage of the MOC of a component parameter. For example, if a Smoke Analysis simulation with a 100% derating factor were to be completed, the Cadence software would compare a component's level of performance directly to the MOC. If a Smoke Analysis simulation with an 80% derating factor were to be completed, the Cadence software would compare a component's level of performance to 80% of the MOC. Designing a circuit with components that operate well under the MOC will certainly have a greater life expectancy than a circuit with components that operate very close to or at the MOC.

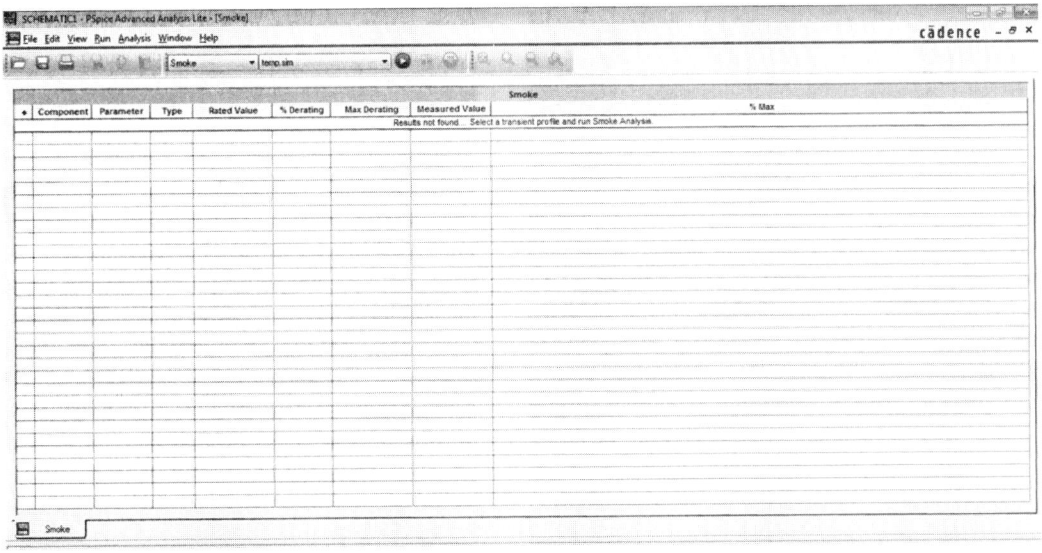

Figure 12.34 — AMS ADVANCED ANALYSIS window, Smoke.

The *PSpice* or *Allegro AMS Advanced Analysis Simulator* calculates the performance of a component based on certain parameters. Some of these Smoke Parameters, specifically the ones used in this example, are listed in **Table 12.1** along with their associated components.

The Smoke Parameter abbreviations listed in the left column of Table 12.1 will appear in a PARAMETER column upon simulation and describe the Smoke Analysis type for each component. Before running the Smoke Analysis, it is important to select the derating factor. The default derating setting for the Smoke Analysis simulation is NO DERATING, which implies a 100% derating factor. This derating factor can be set by opening the PROFILE SETTINGS window in the EDIT menu and clicking the SMOKE tab (see **Figure 12.35**).

After closing the PROFILE SETTINGS window, run the simulation. The results for the Smoke Analysis are shown in **Figure 12.36** and share

Table 12.1 ───
Smoke Parameters

Smoke Parameter	Component	Definition (w/unit)
IB	BJT	Maximum base current (A)
IC	BJT	Maximum collector current (A)
PDM	BJT	Maximum power dissipation (W)
RCA	BJT	Thermal resistance, Case-to-Ambient (degC/W)
RJC	BJT	Thermal resistance, Junction-to-Case (degC/W)
SBINT	BJT	Secondary breakdown intercept (A)
SBMIN	BJT	Derated percent at TJ (secondary breakdown)
SBSLP	BJT	Secondary breakdown slope
SBTSLP	BJT	Temperature derating slope (secondary breakdown)
TJ	BJT	Maximum junction temperature (degC)
VCB	BJT	Maximum collector-base voltage (V)
VCE	BJT	Maximum collector-emitter voltage (V)
VEB	BJT	Maximum emitter-base voltage (V)
CI	Capacitor	Maximum ripple (I)
CV	Capacitor	Voltage rating (V)
SLP	Capacitor	Temperature derating slope (V/degC)
TBRK	Capacitor	Breakpoint temperature (degC)
TMAX	Capacitor	Maximum temperature (degC)
IV	Current Supply	Maximum voltage current source can withstand (V)
LI	Inductor	Current rating (I)
LV	Inductor	Dielectric strength (V)
PDM	Resistor	Maximum power dissipation (W)
RBA	Resistor	Slope of power dissipation vs. temperature (W/degC)
RV	Resistor	Voltage rating (V)
TMAX,TB	Resistor	Maximum temperature resistor can withstand (degC)
VI	Voltage Supply	Maximum current voltage source can withstand (I)

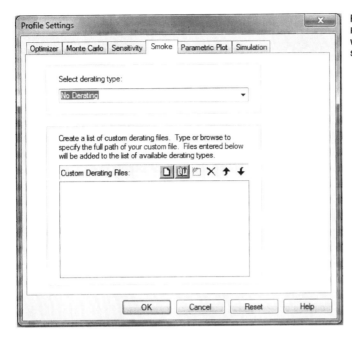

Figure 12.35 —
PROFILE SETTINGS
window, NO DERATING
setting.

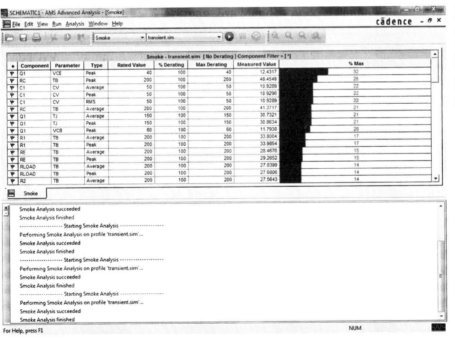

Figure
12.36 — AMS
ADVANCED
ANALYSIS
window,
Smoke
Analysis, NO
DERATING
setting.

a similar layout and display to those of the Sensitivity Analysis. The components are listed in the first column; the second column names the smoke parameter tested. These results can be given in three representations, using a peak value, average value, or RMS value. The type of measurement is indicated in the third column. The rated value represents the MOC for the particular component. Notice the % DERATING column is constant at 100 since the default setting for the simulation is NO DERATING. The % MAX column describes how close in percentage the component operates within its maximum value. This percentage is calculated by taking the value from the MEASURED VALUE column and dividing by the rated value and multiplying by 100. The results are sorted by the % MAX value in descending order. According to the simulation results, all the components fall well under the 100% derating factor (MOC).

The Smoke Analysis simulation will be done once more, but this time it will be completed with standard derating factors. Open the PROFILE

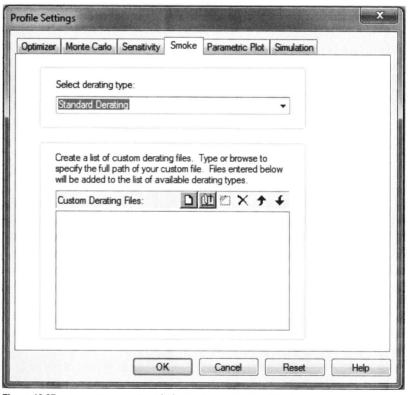

Figure 12.37 — PROFILE SETTINGS window, STANDARD DERATING setting.

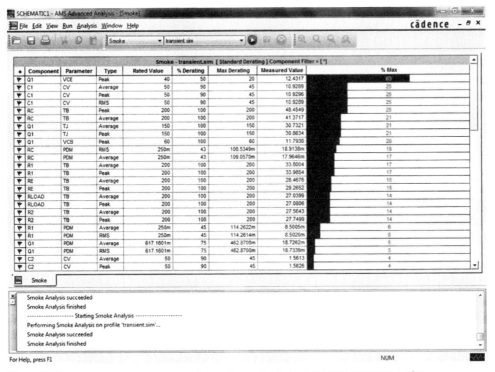

	Component	Parameter	Type	Rated Value	% Derating	Max Derating	Measured Value	% Max
▼	Q1	VCE	Peak	40	50	20	12.4317	63
▼	C1	CV	Average	50	90	45	10.9289	25
▼	C1	CV	Peak	50	90	45	10.9296	25
▼	C1	CV	RMS	50	90	45	10.9289	25
▼	RC	TB	Peak	200	100	200	48.4549	25
▼	RC	TB	Average	200	100	200	41.3717	21
▼	Q1	TJ	Average	150	100	150	30.7321	21
▼	Q1	TJ	Peak	150	100	150	30.8834	21
▼	Q1	VCB	Peak	60	100	60	11.7930	20
▼	RC	PDM	RMS	250m	43	108.5349m	18.9138m	18
▼	RC	PDM	Average	250m	43	109.0570m	17.9646m	17
▼	R1	TB	Average	200	100	200	33.8004	17
▼	R1	TB	Peak	200	100	200	33.9854	17
▼	RE	TB	Average	200	100	200	28.4676	15
▼	RE	TB	Peak	200	100	200	29.2652	15
▼	RLOAD	TB	Average	200	100	200	27.0399	14
▼	RLOAD	TB	Peak	200	100	200	27.0806	14
▼	R2	TB	Average	200	100	200	27.5643	14
▼	R2	TB	Peak	200	100	200	27.7499	14
▼	R1	PDM	Average	250m	45	114.2622m	8.5005m	8
▼	R1	PDM	RMS	250m	45	114.2614m	8.5020m	8
▼	Q1	PDM	Average	617.1601m	75	462.8700m	18.7262m	5
▼	Q1	PDM	RMS	617.1601m	75	462.8700m	18.7339m	5
▼	C2	CV	Average	50	90	45	1.5613	4
▼	C2	CV	Peak	50	90	45	1.5626	4

Figure 12.38 — AMS ADVANCED ANALYSIS window, Smoke Analysis, STANDARD DERATING setting.

SETTINGS window for the Smoke Analysis and select STANDARD DERATING (see **Figure 12.37**). This will implement a set of derating factors designated within the Cadence software for a given component. These two derating types are the only selectable choices in the PROFILE SETTINGS window, though it is possible for a user to declare derating factors in custom derating files and import them.

After making this change to the profile, close the PROFILE SETTINGS window and run the simulation. The results are shown in **Figure 12.38**. Take note of the % DERATING column. The column is no longer fixed at 100. To understand how this will impact the result, we will dissect the first row, Q1. The VCE (max collector emitter voltage) smoke parameter is rated at 40 V. The derating factor imposed by the Cadence software is 50. This means that the derated maximum value will be 50% of the rated value, 20 V. This is indicated in the MAX DERATING column. The % MAX will be calculated based on the measured collector-emitter voltage with respect to the MAX DERATING value, 20. This comes out to approximately 63%. With the results again listed in descending order in the % MAX column, and based on these results with standard derating, all components still fall within an acceptable region of operation.

In the event that a % MAX exceeds a rated/derated MOC, the bar in the % MAX column will appear red. If the % MAX value is calculated to be within 90% to 100% of the derated MOC, the bar in the % MAX column will appear yellow.

In Summation

There are two other types of AA not discussed in this chapter. They are the Optimizer and the Parametric Plotter, and they are typically used heavily during the design process of a circuit. The Optimizer can assist in finding certain component values to ensure that a design will meet a user-defined specification. The Parametric Plotter is a simulation type that sweeps multiple component values or model parameters simultaneously. More information and literature regarding these simulation types can be obtained through Cadence Design Systems.

Appendix

Additional Circuit Examples

This appendix provides additional circuitry that was not covered in earlier chapters. In these pages we are going to look at circuits for 1) Low-Pass Filter; 2) Operational Amplifier; 3) Phase-Shift Oscillator, 4) Astable Multivibrator; and 5) Hartley Oscillator.

1) Low-Pass Filter

A low-pass filter is an electronic filter that passes low-frequency signals and attenuates (reduces the amplitude of) signals with frequencies higher than the cutoff frequency. The actual amount of attenuation for each frequency varies from filter to filter. It is sometimes called a high-cut or treble cut filter when used in audio applications. As one would imagine, a low-pass filter is the opposite of a high-pass filter, and a band-pass filter is a combination of a low-pass and a high-pass. See the low-pass RC circuit in **Figure A.1**. The AC Analysis in **Figure A.2** shows VIN versus VOUT from

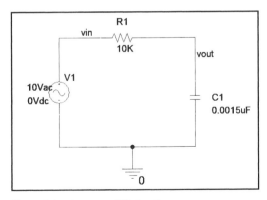

Figure A.1 — Low-pass RC circuit.

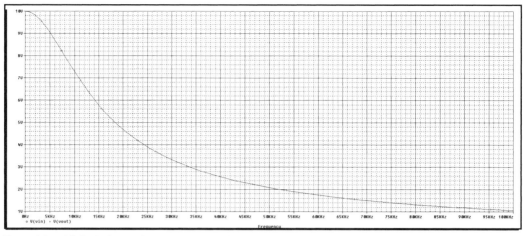

Figure A.2 — AC Analysis showing VIN versus OUT for the low-pass RC filter.

1 Hz to 100 kHz. Other output graphs could be phase versus frequency, linear amplitude versus frequency, and logarithmic amplitude versus frequency.

2) Operational Amplifier

In **Figure A.3** an operational amplifier has been introduced as a *PSpice* element, providing gain for an active filter. An AC Analysis is shown. **Figure A.4** offers a Bode plot of amplitude versus frequency, with results showing VOUT/VIN. It has range of frequencies from 1 Hz to 10 kHz.

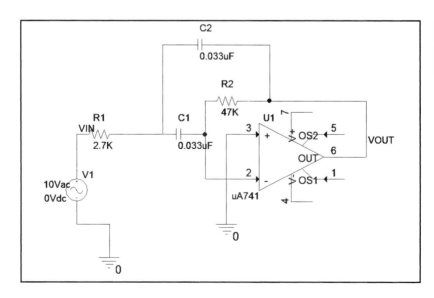

Figure A.3 — Operational amplifier as a *PSpice* element.

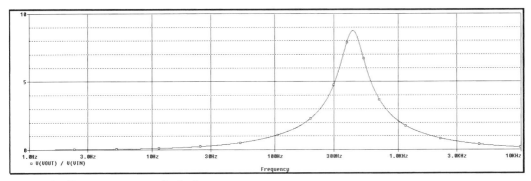

Figure A.4 — Frequency response, Bode plot.

3) Phase-Shift Oscillator

The RC phase-shift oscillator circuit seen in **Figure A.5** is a great example of using a common op-amp to make a sinusoidal oscillator. In the original circuit, R5 was a 1 MΩ potentiometer. A 300 kΩ fixed resistor can be used, giving a gain of –300 k/10 k = –30. V1 was added to kickstart the oscillator for the simulation. Since the RC-RC-RC feedback network has a

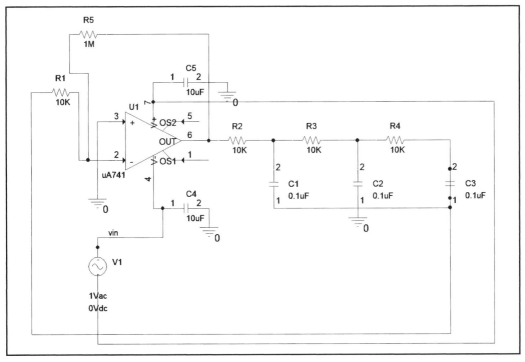

Figure A.5 — RC phase-shift oscillator circuit.

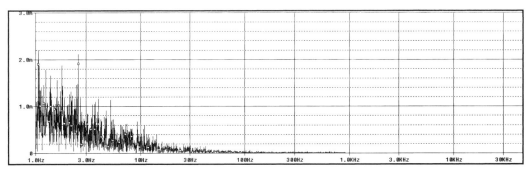

Figure A.6 — Output of circuit, oscillations.

gain of $-\frac{1}{29}$ at the frequency where oscillation will occur, the output graph of pin 6 versus time (a Transient Analysis) will show the growth of oscillation until clipping occurs. **Figure A.6** shows the output of the circuit, and the oscillations can be observed. Maximum variation can be seen between 1 Hz and 3 Hz. As frequency increases, variation in signal decreases. Readers may try decreasing the value of R5 and rerunning the analysis. Eventually the oscillator won't start due to insufficient gain.

4) Astable Multivibrator

Figure A.7 show an astable multivibrator. Again, V1 was added to kickstart the circuit. We implemented a Transient Analysis, evaluating the value of resistance and capacitance to calculate time. Initially, the input voltage was set to 12 V, then by reducing voltage we observe that time is increasing as voltage decreases (see **Figure A.8**).

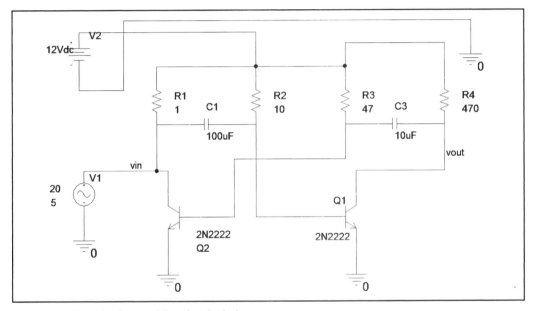

Figure A.7 — Circuit implemented, Transient Analysis.

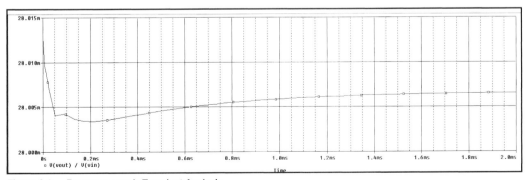

Figure A.8 — Response graph, Transient Analysis.

5) Hartley Oscillator

Figure A.9 shows a Hartley oscillator circuit, which works from frequencies below the radio range up into the gigahertz range. L1 is tapped, forming the reactive divider characteristics of a Hartley circuit. C4 and L1 form a resonant tank circuit to determine the oscillator frequency. The amplifier is a 2N2222 BJT. In **Figure A.10** we have a buffer amplifier that could be added to the oscillator output. V2 is added to simulate the output of the oscillator that would drive the buffer amplifer. A buffer circuit provides isolation; decoupling prevents ac signal from flowing between circuits. Gain bandwidth product is a measure of a transistor's ability to amplify a high-frequency signal (see **Figure A.11**).

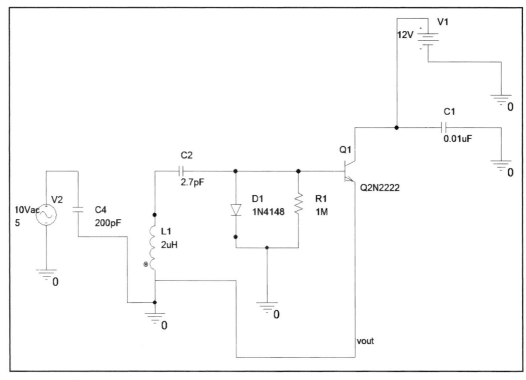

Figure A.9 — Circuit of Hartley oscillator.

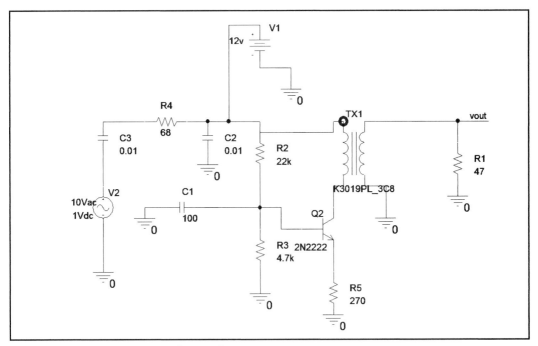

Figure A.10 — Buffer circuit added to RF amplifier.

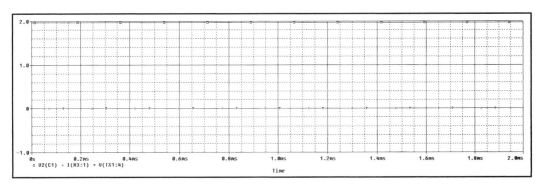

Figure A.11 — Voltage, current response.

FEEDBACK

Please use this form to give us your comments on this book and what you'd like to see in future editions, or e-mail us at **pubsfdbk@arrl.org** (publications feedback). If you use e-mail, please include your name, call, e-mail address and the book title, edition and printing in the body of your message. Also indicate whether or not you are an ARRL member.

Where did you purchase this book?   ☐ From ARRL directly   ☐ From an ARRL dealer

Is there a dealer who carries ARRL publications within:

☐ 5 miles   ☐ 15 miles   ☐ 30 miles   of your location?   ☐ Not sure.

License class:

☐ Novice   ☐ Technician   ☐ Technician with code   ☐ General   ☐ Advanced   ☐ Amateur Extra

Name _____   ARRL member?   ☐ Yes ☐ No

_____   Call Sign _____

Address _____

City, State/Province, ZIP/Postal Code _____

Daytime Phone   (   ) _____   Age _____

If licensed, how long? _____

Other hobbies _____   E-mail _____

_____

Occupation _____

For ARRL use only	CSA
Edition	1 2 3 4 5 6 7 8 9 10 11 12
Printing	1 2 3 4 5 6 7 8 9 10 11 12